Michael M. Resch • Xin Wang • Wolfgang Bez
Erich Focht • Hiroaki Kobayashi
Editors

Sustained Simulation Performance 2012

Proceedings of the joint Workshop on High Performance Computing on Vector Systems, Stuttgart (HLRS), and Workshop on Sustained Simulation Performance, Tohoku University, 2012

Springer

Editors
Michael M. Resch
Xin Wang
High Performance Computing Center
Stuttgart (HLRS)
University of Stuttgart
Stuttgart
Germany

Wolfgang Bez
NEC High Performance Computing
Europe GmbH
Düsseldorf
Germany

Erich Focht
NEC High Performance Computing
Europe GmbH
Stuttgart
Germany

Hiroaki Kobayashi
Cyberscience Center
Tohoku University
Sendai
Japan

Front cover figure: Flow simulation of a F1 model by Building Cube Method. Illustration by Cyberscience Center, Tohoku University, 6-3 Aramaki-aza-aoba, Aoba, Sendai 980-8578, Japan.

ISBN 978-3-642-32453-6 ISBN 978-3-642-32454-3 (eBook)
DOI 10.1007/978-3-642-32454-3
Springer Heidelberg New York Dordrecht London

Library of Congress Control Number: 2012948969

Mathematics Subject Classification (2010): 68Wxx, 68W10, 68Mxx, 68U20, 76-XX, 86A10, 70FXX, 92Cxx

Printed on acid-free paper

Springer is part of Springer Science+Business Media (www.springer.com)

Sustained Simulation Performance 2012

Preface

Sustained simulation performance is a widely ignored issue in high-performance computing. Typically the focus is on peak performance and on Linpack results. Rarely we do see a discussion about real-world applications and their performance on large-scale systems.

This book presents the results of the 14th Teraflop Workshop which was hosted by the High Performance Computing Center Stuttgart/Höchstleistungsrechenzentrum Stuttgart (HLRS) in December 2011 as well as the contributions of the 15th workshop in this series. In order to adapt the title of the workshop series to the changing technology in high-performance computing the workshop was renamed to "Workshop on Sustained Simulation Performance." It was held in March 2012 at the Tohoku University in Sendai, Japan.

This book contains contributions focused on the issue of sustainable performance on large-scale systems. This includes issues of exaflop computing as one important part. The three other parts are focusing on real-world applications and their performance on state-of-the-art HPC architectures. These contributions together give an overview of the level of performance that is available for the scientific community today.

The workshop series is based on a project that was initiated in 2004. The Teraflop Workbench Project was founded initially as a collaboration between the High Performance Computing Center Stuttgart (HLRS) and NEC Deutschland GmbH (NEC HPCE) to support users to achieve their research goals using high performance computing.

Since then a series of workshops have put their focus on sustainable performance. These workshops have become a meeting platform for scientists, application developers, international experts, and hardware designers to discuss the current state and future directions of supercomputing with the aim of achieving the highest sustained application performance.

Work in the Teraflop Workbench project gives us insight into the applications and requirements for current and future HPC systems. We observe the emergence of multi-scale and multi-physics applications, the increase in interdisciplinary tasks, and the growing tendency to use today's stand-alone application codes as modules in prospective, more complex coupled simulations. At the same time, we notice the current lack of support for those applications. Our goal is to offer an environment that allows users to concentrate on their area of expertise without spending too much time on computer science itself.

The first stage of the Teraflop Workbench project (2004–2008) concentrated on user's applications and their optimization for the 72-node NEC SX-8 installation at HLRS. During this stage, numerous individual codes, developed and maintained by researchers or commercial organizations, have been analyzed and optimized. Several of the codes have shown the ability to outreach the TFlop/s threshold of sustained performance. This created the possibility for new science and a deeper understanding of the underlying physics.

The second stage of the Teraflop Workbench project (2008–2012) focuses on current and future trends of hardware and software developments. We observe a strong tendency towards heterogeneous environments at the hardware level. At the same time, applications become increasingly heterogeneous by including multi-physics or multi-scale effects. The goal of the current studies of the Teraflop Workbench is to gain insight into the developments of both components. The overall target is to help scientists to run their applications in the most efficient and most convenient way on the hardware best suited for their purposes.

We would like to thank all the contributors of this book and the Teraflop Workbench project. We thank especially Prof. Hiroaki Kobayashi for the close collaboration over the past years and are looking forward to intensifying our cooperation in the future.

Stuttgart, June 2012

Michael M. Resch
Xin Wang
Uwe Küster

Contents

Part I
Exascale Computing: New Challenges in Software and Hardware

Beyond Exaflop Computing: Reaching the Frontiers of High Performance Computing

Michael M. Resch

Abstract High Performance Computing (HPC) has over the last years benefitted from a continuous increase in speed of processors and systems. Over time we have reached Megaflops, Gigaflops, Teraflops, and finally in 2010 Petaflops. The next step in the ongoing race for speed is the Exaflop. In the US and in Japan plans are made for systems that are supposed to reach one Exaflop. The timing is not yet clear but estimates are that sometime between 2018 and 2020 such a system might be available. While we debate how and when to install an Exaflop system discussions have started about what we have to expect beyond Exaflops. There is a growing group of people who have a pessimistic view on High Performance Computing assuming that the continuous development might come to an end. However, we should have a more pragmatic view. Facing a change in hardware development should not be seen as an excuse to ignore the potential for improvement in software.

1 Introduction

In 1893 Frederick Jackson Turner wrote an essay on the significance of the frontier in American history [1]. Referring to a bulletin of the Superintendent of the Census for 1890 he found that the impressive move westwards of the US-American settlers had come to an end. In his description of the advance of the frontier Turner identifies five barriers that were reached over time: the Alleghenies, the Mississippi, the Missouri, the Rocky Mountains, and finally the Pacific Ocean. With the advent of the settlers at the Pacific Ocean, Turner argues, the development of the USA turned mostly inwards and focused on the development of the settled country.

M.M. Resch (✉)
University of Stuttgart, Höchstleistungsrechenzentrum Stuttgart (HLRS), Nobelstrasse 19, 70569 Stuttgart, Germany
e-mail: resch@hlrs.de

M. Resch et al. (eds.), *Sustained Simulation Performance 2012*,
DOI 10.1007/978-3-642-32454-3_1, © Springer-Verlag Berlin Heidelberg 2013

In Supercomputing we have seen a similar breath taking advance. Over only a few decades we have reached Megaflops, Gigaflops, Teraflops, and Petaflops and are approaching Exaflops. While many prepare for the usage of such Exascale systems, others start to doubt whether we will be able to reach Exaflops or go beyond that barrier. From a technical point of view there is no doubt that Exaflops are possible. The driving factor, however, is no longer innovation but rather a massive usage of standard parts and components. Quantitative growth has replaced qualitative improvement.

In this paper we have a look at how we got to the situation in which we are in High Performance Computing today. We will then investigate the ongoing trends and extrapolate future developments. Instead of arguing for new efforts to build faster systems we will emphasize that software is the key to solutions in simulation. Hence, we will argue that the improvement of software is much more important than further heroic efforts in designing ever faster hardware.

2 History

High Performance Computing has seen a long history of progress over the last six decades. For a long time Moore's assumption about the doubling of transistors on a die [2] and the corollary of the doubling of speed every 18 months turned out to be right. When the increase in clock frequency started to level out, parallelism became the driving factor. Parallelism came so far in two waves. The first one started in the late 1980s and by the early 1990s created a number of interesting concepts. However, most of the advanced concepts—with sometimes very large numbers of processors—failed, and a number of companies developing such systems disappeared from the High Performance Computing arena pretty fast.

2.1 Architecture in Review

High Performance Computing grew out of the primeval soup of computing. Historical accounts [3] claim that Seymour Cray at a certain point in time decided he wanted to build the fastest systems in the world, rather than economically interesting ones. Technically Seymour Cray was following some basic principles which we still have to consider today. For example, the round shape of his first systems was owed to an attempt to reduce distances. The further success of his first company proofed that speed and economic success were possible at the same time. His failure with follow-on projects showed that things did change after a while. As long as the computer was a special purpose instrument for a limited number of users High Performance Computing was based on special purpose systems with special prices.

The advent of the personal computer started to change things. The computer became a ubiquitous instrument. Budgets for computer hardware were gradually moving from special purpose systems to general purpose hardware. Already in 1990 Eugene Brooks coined the term "Killermicros" [4], describing the end of what was considered to be "dinosaur" systems, and their replacement by microprocessor based systems. It did not take too long until Brooks was proven right. The number of systems using specialized processor technology started to dwindle away and by the year 2000 only small "game reserves" were left. By the year 2009 such specialized processors were virtually extinct when NEC announced its withdrawal from the Japanese Next Generation Supercomputing Project. However, recent announcements of NEC suggest that we may see another round of specialized processors in the future. The driving factor behind such specialized processors is the need of applications.

It is interesting to see that the "killermicros" did not only make an end to specialized processors in High Performance Computing. They also started to cannibalize each other. Over a period of about 10 years a number of processor architectures disappeared from the TOP 500 list [5]. While in the year 2000 the TOP 500 presented five different processor architectures with a substantial share of systems, that same list of the year 2012 shows that about 90 % of the systems are based on the x86-architecture. The main non-x86 architecture is the IBM Power architecture which is used in various IBM BlueGene [6] systems.

The trend in microprocessor architectures was accompanied by a trend in system architectures. In 1994 Donald Becker and Thomas Sterling started what they called the "Beowulf Project" [7]. As a result of this project—that was building a High Performance Computing cluster from standard components—clusters became extremely popular in High Performance Computing. In 2000 expectations were high that future High Performance Computing systems would all be clusters based on "Components Off The Shelf" (COTS). To some extent this has become true. Over 80 % of the systems listed in the TOP 500 list in June 2012 actually are clusters.

The situation as described already shows that a small number of trends shape the landscape of High Performance Computing. After this short historical review we now have a look at the current situation and try to estimate the future developments in hardware.

3 Situation

The result of the first wave of parallelism was a number of systems that typically provided a moderate number of processors in a single system. For a while the largest systems were hovering around 1,000 and up to 10,000 processors or cores. The fastest system in the world in 1993—as presented by the TOP 500 list [5]—was based on 1024 processors. The fastest system in 2003 was based on 5120 processors. The increase was a factor of about five. We have to consider though that in 2003 the number one system—the Japanese Earth Simulator from NEC—was using special fast vector processors, which allowed it to provide a relatively high peak performance with a relatively low number of processors. But even if we look at the top ten systems of the list, we only find an increase in the level of parallelism of about ten—from 340 processors per system in 1993 to 4180 processors per system in 2003. The parallelism of this kind was not easy to master but message passing was a good approach, and software programmers could easily keep track of their thousands of processes.

Since about 2003 we have seen a second phase of parallelism. This was partially driven by the IBM BlueGene project [6] which was using a larger number of slower processors. Over the last 2–3 years graphics processing units (GPUs)—with their hundreds of cores on one card—have further enhanced the trend. The currently valid TOP 500 list of June 2012 shows a system with more than 1.500.000 cores at the top. This is a factor of 300 over the last 9 years. Looking at the top ten systems we see a factor of 100 over the last 9 years—from 4180 cores per system in 2003 to 418.947 cores per system in 2012. So, while we had 10 years to adapt to an increase in number of processors of ten from 1993 to 2003 we now had 9 years to adapt to a factor of 100 from 2003 till 2012. From a programmer's point of view things are getting out of control.

3.1 Processors

The basis for the top systems in High Performance Computing are currently many-core processors. The number one system in 2011—the Japanese K-Computer—relied on the Fujitsu SPARC64 VIIIfx, a many-core processor with 8 cores [8]. More than 88.000 of these processors are bundled. Other large scale systems are based on the AMD Opteron 6200 processor with 16 cores. In both cases the clock frequency is comparably low. It is 2 GHz for the SPARC processor and 2.3 GHz for the AMD processor.

Another standard building block used in very large scale systems is the so called general purpose graphics processing unit (GPGPU). Based on e.g. NVIDIA 2050 cards [9] a high level of peak performance as well as of Linpack performance is made possible. The NVIDIA 2050 comes with 448 cores which again increases the number of cores for the user.

The background of this increase in number of cores is clear. The International Technology Roadmap for Semiconductors [10] indicates that what was suggested by G.E. Moore more than 50 years ago is still valid. The feature size is shrinking and it will keep doing so for a number of years to come. While we cannot increase the clock frequency anymore—basically because of the high leakage that comes with high clock frequencies—we still can substantially increase the number of transistors on a single chip. As a consequence, we are increasing the number of cores on a chip instead of shrinking the chip and increasing the frequency.

The SPARC VIIIfx and the NVIDIA 2050 find themselves on two ends of a spectrum defined in terms of complexity of cores and number of cores. The SPARC VIIIfx processor comes with a rather complex core design. Each of the cores could be described as "fat and fast". The typical graphic cards like the NVIDIA 2050 come with a very large number of cores. These cores can be described as "slim and slow".

Finally, we should mention some details about a special purpose system from IBM. The IBM BlueGene [6] is an architecture that is currently based on the IBM Power PC A2 processor. It comes with a relatively low clock frequency of 1.6 GHz and has 18 cores of which 16 are used for computing. The processor's architecture is interesting in that it provides one extra core for running an operating system and one extra core as a spare core in case one of the 16 computing cores fails. These two strategies—operating system offloading and hardware support for fault tolerance—are increasingly becoming important. Unfortunately the Power PC A2 is only available in the BlueGene system for High Performance Computing. Its market share will hence be relatively small. It remains to be seen how this architecture will further evolve. Technically the processor can be considered to be closer to the "fat and fast" solution than to the "slim and slow" solution.

3.2 *Networks*

In the field of internal communication networks we have seen a variety of solutions in the past [11]. For a while several solutions were competing in the field of cluster computing. With the advent of Infiniband [12] the situation has changed. The new technology has practically replaced all other special solutions in the cluster market. Of the 50 fastest systems in the TOP 500 23 are clusters based on Infiniband. The interesting finding is that 26 of the TOP 50 system are using some kind of proprietary network. Only one system is still based on Gigabit Ethernet (ranked number 42 in November 2011).

When we turn the pages of the TOP 500 list we find that starting around a ranking of 150 the number of Gigabit Ethernet installations substantially increases. This indicates that for high-end systems Infiniband is a must. This is also supported when looking at the level of sustained Linpack performance in the list. The typical Gigabit Ethernet system achieves about 50–60 % of peak performance for the Linpack benchmark. For an Infiniband system this ratio is typically in the range between 45 % and 85 %. The big variation indicates that Infiniband is used to build a large variety of different network topologies.

3.3 *Architectures*

Looking at architectures we find that clusters dominate the TOP 500 list. About 80 % of the fastest systems in the world are in that group. The rest of about 20 % is based on an MPP architecture approach. Even though clusters are the biggest group we look at the MPP architectures. They seem to be outdated but keep a constant share of about 20 % over the last 9 years, while other types of architectures have disappeared. What is more interesting: in terms of performance MPPs have a much larger share of the TOP 500 list—in the range of 40 %. This is because MPPs can typically be found in the upper part of the TOP 500. So, when talking about real High Performance Computers we find that MPP and clusters are two competing technologies at equal footing.

One of the reasons for a renaissance of the MPPs is the IBM BlueGene architecture. Originally the concept was based on a relatively light-weight processor. The new Blue Gene/Q has a relatively strong processor but comes with a lower clock frequency than comparable systems. The network is proprietary and provides a 5D-Torus.

In general, one of the main features of MPPs systems seems to be the better network connectivity. The basic performance numbers (latency and bandwidth) for MPI are typically comparable to what Infiniband can offer. However, the better network connectivity should increase the level of sustained performance. Analysing the fastest 50 systems in the world in November 2011, we see that proprietary interconnects on average achieve 78 % of sustained performance for the Linpack benchmark. Infiniband based systems achieve about 74 %. This is not a big difference. One may wonder whether this is the reason why Cray decided to give up its proprietary network development in 2012 [13].

A further investigation of the list shows that low sustained performance is caused by the usage of graphics cards. Such cards provide a high level of peak performance but typically do not work well in terms of sustained Linpack performance. The average sustained Linpack performance of the top 50 systems is 76 %. The average of the six systems that make use of NVIDIA cards is 51 % only. This is a clear indication that such systems do not show satisfactory sustained performance for classical High Performance Computing applications.

What is further interesting is the evaluation of network architectures when we eliminate the NVIDIA results. Without these systems both Infiniband based systems and proprietary systems show an average of 80 % of Linpack performance. The maximum for proprietary networks is 93 %, for Infiniband it is 89 %. The minimum for proprietary networks is 72 %, and for Infiniband it is 59 %. However, the relatively low minimum for Infiniband is the exception to the rule.

4 Potential Paths Forward

The last 20 years have shown that changes in technology in High Performance Computing happen all the time and that predictions are difficult to make. However, there are a number of findings.

The number of cores in a High Performance Computer will further increase. There is currently no way of avoiding a situation in which millions of cores form the compute backbone of a High Performance Computer system. We may see solutions where the large number of cores is hidden from the user. However, this is certainly going to happen based on some kind of software solution. Most likely we are going to see compilers that support a high level of parallelism in a single node—whatever the term "node" is going to mean in the coming years.

Given the actual lack of advantage for proprietary networks one has to expect that Infiniband—or a follow-on technology—is going to gain more ground also in the TOP 500 list. Economic considerations might lead to an end of proprietary network development in much the same way that they have caused a substantial reduction in processor architectures available for High Performance Computing.

4.1 Exaflops

The analysis of technologies for High Performance Computing shows that an Exaflop system can be built but will be extremely difficult to handle. Furthermore the costs for operating such a system will be relatively high. Recent estimates expect a power consumption in the range of 20 up to 75 MW. Even though this may technically be possible it may not make sense financially. A discussion has hence started about the feasibility of Exaflop computing. While some argue for a change of our programming style [14] for Exaflop computing others discuss whether we will ever see such systems [15].

5 Discussion

An investigation of the current technology available for High Performance Computing systems reveals a number of findings that may be helpful for a future strategy in HPC. First and foremost we can safely assume that we will see a further shrinking of feature-size for semiconductors. As a consequence there is still room for improvements of processors. The same cannot be said for clock frequencies. We will have to live with clock rates of a low single digit number of GHz for the coming years. As a result massive parallelism is the only option that we have to increase speed. Assuming that the trend of the last 10 years will continue we may expect to see systems with a billion cores in 8–10 years from now.

Such a number may be prohibitively large since it may lead to power consumptions beyond the financial reach of typical HPC centers. On the other hand our existing programming models will be challenged by such architectures to an extent that may lead to severe problems for programmers in HPC. MPI was not designed for such large numbers of processes. Whether the concept can be adapted has to be seen. OpenMP has proven to work well for a small number of shared memory processes—in the order of 8–16. However, it cannot be considered to be the method of choice for shared memory systems with thousands of cores. So, although we may get a growth in speed of systems, we have to accept that the time is over when supercomputers regularly provided such an increase in speed at a relatively low cost.

However, the situation has a number of positive aspects for those who are willing to change. First, we need to understand that further technical improvement should not be sought in the increase of speed of a single processor. We need to shift our attention from the processor to the memory. Investigations show that for many applications the slow memory is the limiting factor. Hence, we should investigate options to shift power budgets from processor cores to memory. Recent investigations show that such an open investigation of shifting the focus of attention might be extremely beneficial for certain applications [16].

The key aspect of technological development, however, is the end of Moore's law. We still see some increase in number of transistors per square inch but the race is pretty much over. A technology is reaching the end of the line. After about 60 years of technical development we have to accept that we have to change the way we think and work when it comes to supercomputing. Much as the United States turned towards internal development after having reached the Pacific Ocean, the High Performance Computing community should turn inwards. There it will find a number of problems that still wait for a solution.

5.1 Hardware Issues

The focus of hardware development over the last decades was mainly on speed of processors. Hence, when power issues came up the target of hardware designers was not an improvement of the speed or quality of memory subsystems. Instead they turned to multi-core solutions. For most applications this does not provide a lot of improvement. However, nominally systems get faster if the number of cores is increased.

Little progress has also been made in networks. Standardization has basically eliminated competition and the relative speed of interconnects compared to compute speed has been shrinking over the last decades consistently.

5.2 Software Issues

Programming models have not seen much innovation once MPI and OpenMP where standardized. Although MPI is constantly enhanced it cannot hide the fact that it was designed with only thousands of processes in mind. The growing number of processes in the range of millions and beyond is not reflected in this model.

Another key problem for software is the software life cycle. While hardware concepts constantly change and old hardware is replaced after about 3–5 years software typically has a life span of about 20 years. So once software has matured the corresponding hardware concepts already have become obsolete. The fact that we can expect hardware development to slow down is a chance for software developers to come up with solutions which are better synchronized with hardware in the future.

5.3 *Modeling Issues*

Over the last decades there was a small group of scientists who have opted for better models in simulation. This community has been basically silenced by the ever increasing speed of High Performance Computers. The fact that much larger models could be simulated has substantially contributed to a slow-down in model development. Publications and scientific visibility could easily be achieved by using larger models and pointing at the new features of the simulation that could only be made visible by using faster systems. The fact that speed may not be the key factor in hardware development anymore may give modeling experts the time needed to show the potential of new models and further model improvement.

6 Summary

The current technical trends in High Performance Computing hint at an end of the super linear increase of speed that we have seen over the last six decades. To adapt to this situation centers and users alike will have to adapt their strategies. By no means, however, should this be considered to be only a problem for the community. It should be rather taken as a chance to pick up the many issues and problems that were ignored in the past because they were overshadowed by the stunning race for ever faster systems. Both in terms of hardware and in terms of software this development offers new opportunities and presents new and interesting research questions which deserve our full attention. Those centers and users who take this development as a chance to improve their work and exploit the technology to its fullest will not only survive the end of Moore's law but will continue to push the frontiers of simulation.

Acknowledgements The author would like to thank Hans Meuer and his team for providing extremely valuable insight into the development of High Performance Computing over the last 20 years by collecting information in the TOP 500 list.

References

1. Frederic Jackson Turner: The Significance of the Frontier in American History. Penguin Books (2008)
2. G.E. Moore: Cramming more components onto integrated circuits, Electronics, 38(8), 114–117 (1965)
3. Charles J. Murray: The Supermen – The Story of Seymour Cray and the Technical Wizards behind the Supercomputer, John Wiley & Sons (1997)
4. http://www.websters-online-dictionary.org/definitions/KILLER+MICRO (2012)
5. TOP 500 List: www.top500.org (2012)
6. http://www.research.ibm.com/bluegene/index.html (2012)

7. T. Sterling, D. Savarese, B. Fryxell, K. Olson, D.J. Becker: Communication Overhead for Space Science Applications on the Beowulf Parallel Workstation, Proceedings of High Performance Distributed Computing (1995)
8. Takumi Maruyama, Tsuyoshi Motokurumada, Kuniki Morita, Naozumi Aoki: Past, Present, and Future of SPARC64 Processors, FUJITSU Sci. Tech. J., Vol. 47, No. 2, pp. 130–135 (2011)
9. http://www.nvidia.com/object/personal-supercomputing.html (2012)
10. ITRS: International Technology Roadmap for Semiconductors 2011 Edition http://www.itrs.net/Links/2011ITRS/Home2011.htm
11. Michael M. Resch: Trends in Architectures and Methods for High Performance Computing Simulation, in B.H.V. Topping and P. Ivanyi (Eds.), Parallel Distributed and Grid Computing for Engineering, pp 37–48, Saxe-Coburg Publications, Stirlingshire, Scotland (2009)
12. http://www.infinibandta.org/ (2012)
13. http://investors.cray.com/phoenix.zhtml?c=98390&p=irol-newsArticle&ID=1690642&highlight= (2012)
14. G. R. Liu, On Future Computational Methods for Exascale Computers, iacm expressions, 30, 8–10, December 2011 (2011)
15. Peter Kogge: Next-Generation Supercomputers, IEEE Spectrum, February 2011 (2011)
16. Richard Vuduc: A Theory for Co-Designing Algorithms and Architctures under Power & Chip-Area Constraints, ISC 2012, Hamburg, Germany (2012)

Architectural Considerations for Exascale Supercomputing

Yasuo Ishii

Abstract Towards exascale supercomputing, both academia and industry have started to investigate the future HPC technologies. One of the most difficult challenges is the enhancement of energy efficiency of the computer system. We discuss the energy efficiency of existing architecture in this paper.
With the analysis of the performance of the dense matrix–matrix multiplication (DGEMM), we propose the DGEMM-specialized Vector-SIMD architecture that only requires the small number of processor cores and low memory bandwidth. The DGEMM-specialized Vector-SIMD architecture can outperform existing architectures with respect to several metrics, as far as it is dedicated to limited usages, such as the DGEMM calculation. We conclude that this type of discussion will be essential in designing the future computer architecture.

1 Introduction

The performance of the supercomputer has grown two times per year. As the result of the performance improvement, the world fastest supercomputer has achieved 10 peta floating-point operations per seconds (PFLOPS) in 2011 [2]. The high-performance computing communities initiated the research for the next generation supercomputer, including "exascale supercomputing," having in mind possible approaches toward grand challenge applications on future high-performance computing systems. However, most researchers and developers are also faced with many difficulties in realizing exascale supercomputer [4].

To overcome the difficulties toward exascale supercomputing, the Japanese HPC community identified the following six issues in its technical roadmap [8]: (1) improving energy-efficiency, (2) using memory hierarchy to reduce the traffic

Y. Ishii (✉)
HPC Division, NEC Corporation, 1–10, Nisshin-cho, Fuchu, Tokyo, Japan 183-8501
e-mail: y-ishii@bc.jp.nec.com

M. Resch et al. (eds.), *Sustained Simulation Performance 2012*,
DOI 10.1007/978-3-642-32454-3_2, © Springer-Verlag Berlin Heidelberg 2013

of off-chip memory, (3) exploiting parallelism of millions of processing cores, (4) heterogeneous computing for accelerating the performance, (5) dependability of the millions of processor cores and their interconnections, and (6) productivity for the complicated system. The technical report concluded that the energy-efficiency is the most difficult and essential challenge for the future supercomputing.

To explore the possibility of the exascale supercomputing, we discuss the feature of the existing architectures from the aspect of minimized energy consumption. For quantitative discussion, we use the dense matrix–matrix multiplication (DGEMM) as a benchmark kernel, because important applications are often computationally dominated by DGEMM in such areas as material science. To accelerate the DGEMM operations, utilizing memory hierarchy is essential, because the naive implementation of DGEMM requires about 8.0 bytes per FLOP (B/F) of off-chip memory bandwidth, while modern supercomputers can achieve only about 0.5 B/F. Typically, blocking algorithms are adopted to save the off-chip memory bandwidth, but such algorithms require a storage for the submatrix on the processor chip. Concerning the trade-offs between the off-chip memory bandwidth and the capacity of on-chip memory, many blocking strategies have been proposed. To discuss the trade-offs, we evaluate three widely used architectures: (1) Subword-SIMD, (2) SIMT, and (3) Vector-SIMD. Typically, these architectures have different memory subsystems and different register files, thus requiring different blocking strategies. We demonstrate the energy consumption for the DGEMM and the utilization of the on-chip memory.

Based on the analysis, we propose the DGEMM-specialized Vector-SIMD architecture to maximize the energy-efficiency of the DGEMM. It can reduce hardware resources for the memory subsystem and increase the supported vector length to reduce energy consumption.

2 Dense Matrix–Matrix Multiplication

Dense matrix–matrix multiplication is one of the most important kernel programs of the HPC systems. This kernel program is used in Linpack (which is used in TOP 500 [2, 5]). DGEMM is also utilized in some types of real applications in such a field as nano and material sciences. In this section, we introduce the basic algorithm of DGEMM.

2.1 DGEMM Algorithm

In the BLAS specification, DGEMM is defined as

$$C := \alpha * A * B + \beta * C$$

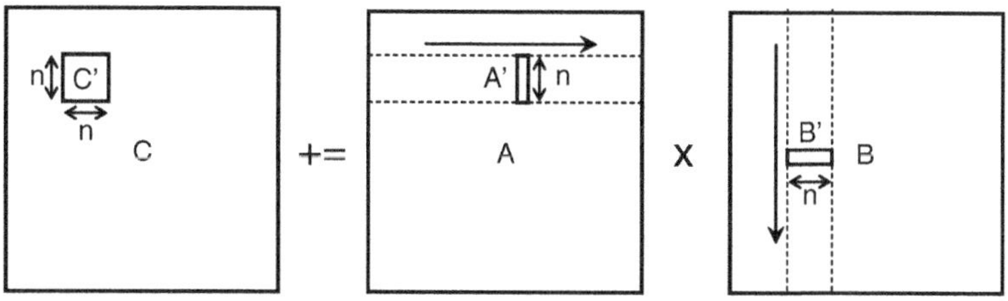

Fig. 1 Blocking strategy of the dense matrix–matrix multiplication

where A, B, and C are appropriate matrices. For simplifying the problem, we use $\alpha = \beta = 1.0$ and we also assume all matrices are square matrices, whose sizes are $n \times n$. One of the simplest implementations of DGEMM is three nested loop. However, it requires a large amount of off-chip memory traffic. The bandwidth requirement of the straightforward approach is 8.0 B/F[1] while modern processors can only realize about 0.5 B/F for the off-chip memory bandwidth. Unfortunately, the reduction of the B/F value is foreseen for the future processors [12].

To improve the computation-efficiency of the DGEMM with such poor memory systems, modern DGEMM libraries including state-of-the-art BLAS library [1] use the blocking algorithm. With the blocking algorithm, matrix C is divided into multiple submatrices (C') that can be stored on the on-chip memory. Each submatrix is computed independently as shown in Fig. 1. When the C' is $n \times n$ submatrix, the requirement of the off-chip memory bandwidth becomes $8.0/n$ B/F, because the multiplication of A' and B' ($16n$ B memory traffic) requires $2n^2$ floating operations. As shown in the example, the blocking algorithm saves the off-chip memory traffic significantly. As a result of the blocking, the sustained performance of DGEMM is expected achieve a close-to-peak sustained performance on many modern processors. In the case of $n = 128$, 0.063 B/F is required to fully utilize the floating operation units.

3 Architecture Design Pattern

Most modern supercomputer systems exploit data-level parallelism, because the instruction-level parallelism has limitations [11]. To utilize the data-level parallelism effectively, many types of the data-level parallelism architectures have been proposed. These architecture designs are already discussed in the existing work [7]. We choose three architecture design patterns, which are widely used for commercial HPC systems, for our analysis (Fig. 2). Three architectures are (A) Subword-SIMD architecture, (B) Single Instruction Multiple-Thread (SIMT) architecture,

[1] Each fused multiply and add operation requires two memory accesses (16B).

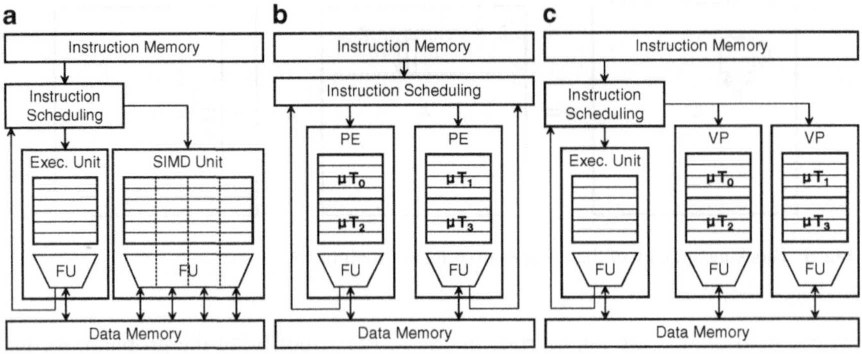

Fig. 2 Architecture design patterns. (*a*) Subword-SIMD, (*b*) SIMT, (*c*) Vector-SIMD

and (C) Vector-SIMD architecture described in this article. We introduce these architectures for further discussions.

3.1 Subword-SIMD

The subword-SIMD processor core employs dedicated data paths and register files which support fixed-length short vector data. The subword-SIMD processor core loads vector data, which are consecutive and aligned on the main memory, and compute them on the dedicated register files. This approach is widely used on the commodity processors, such as x86 SSE and Cell B.E.

The advantage of this architecture is high productivity of the application programs, because the SIMD width is smaller than the other architecture design pattern. This feature is suited in handling a fine-grain data-level parallelism. However, smaller SIMD widths also decrease the performance gain of the vectorization, thus increasing the energy-consumption of the instruction fetching and instruction scheduling.

3.2 SIMT

The SIMT processor core employs multiple processing elements (PEs), which employ own data path and register file. However, the processing element does not have own instruction sequencing unit such as instruction cache and instruction scheduling unit. All processing elements on the core share one instruction sequencing unit to reduce cost and power consumption similarly to classical SIMD processors. Typically, all processing elements execute same instructions at a time, but each processing element also acts as an independent context unlike the classical SIMD approach. This approach is used in the GPU computing.

SIMT effectively utilizes the thread-level parallelism on the SIMD hardware. The massive thread feature helps to avoid the stall time due to the long memory latency, since the processor executes the other threads that do not wait for the memory access. It significantly increases the sustained performance for modern memory systems whose latency achieves hundreds of processor clock cycles. However, the massive thread also decreases the locality of the memory access, because each thread has its own context. It often decreases the efficiency of the on-chip cache and the row-buffer locality of the memory controllers.

3.3 Vector-SIMD

The Vector-SIMD processor core employs one control processor (CP) and multiple virtual processors (VPs), which are typically called vector pipeline. The Vector-SIMD processor core loads any vector data from the main memory, such as consecutive, stride, and pointer chasing. Therefore, Vector-SIMD supports many types of data-level parallelism effectively and realizes high sustained performance for various applications [9]. However, the Vector-SIMD processor core consumes a large amount of resources to support these features effectively, worsening the power-efficiency of the existing HPC systems.

The advantage of the Vector-SIMD is the long vector length. Many HPC applications use very large matrices or vectors to simulate large computational models. For such applications, the energy consumption for instruction fetching and scheduling can significantly be reduced on the Vector-SIMD processor. However, the Vector-SIMD processor cannot work efficiently on the applications that do not include enough data-level parallelism, which tends to confine application fields of Vector-SIMD processors.

4 Blocking Algorithm for Each Architecture

On the DGEMM computation, the processor can easily exploit data-level parallelism because the data structure of DGEMM is vector of vectors. Moreover, the computation of submatrices is easily vectorized too. Modern computer uses some types of vectorization on the computation of the DGEMM.

Subword-SIMD processors compute DGEMM hierarchically because it can exploit data-level parallelism with fine granularity. Typically, 3-level or more hierarchy is used [3, 6, 10]. On the multi-core processors, each core loads own target submatrix into the on-chip memory and computes each submatrix independently even if the last-level cache is shared among multiple cores, as shown in Fig. 3. It simplifies the implementation and reduces the synchronization penalty, but it also

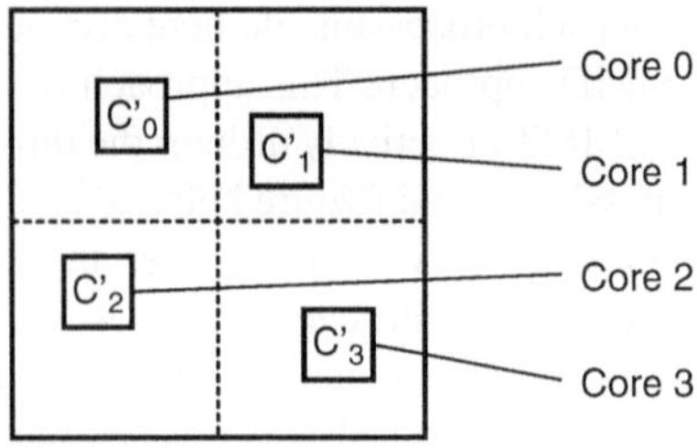

Fig. 3 When multiple cores compute a single matrix in parallel, fixed sized submatrix is allocated to each core. Each core computes the allocated submatrix independently

requires the large amount of on-chip memory because each core requires on-chip memory for own submatrix.

SIMT processors take same approach with Subword-SIMD processors. Typically, the core count of the SIMT processors is larger than that of Subword-SIMD (e.g., current x86 processors have around 8 cores, while the latest GPU employs around 30 cores), indicating that SIMT needs a much more memory bandwidth than Subword-SIMD processors.

On the other hand, the core count of Vector-SIMD is smaller than the other approaches, because each core employs a large number of floating-point arithmetic units. However, the Vector-SIMD architecture does not utilize on-chip memory cache to compute DGEMM. It worsens the energy consumption of Vector-SIMD systems.

Figure 4 shows the relationship between required memory bandwidth and the core count for the processor chip whose peak performance is 10TFLOPS. It shows that the reduction of the number of processor cores contributes to mitigating the requirement of off-chip memory bandwidth. On a 512-core processor with 128 MB on-chip memory, the processor chip has to employ 640 GB/s to supply enough data for fully exploiting the floating-point arithmetic unit, while the 32-core processor with 128 MB on-chip memory requires only 160 GB/s. This results in larger power consumption of the memory system for many-core processors. As shown in Fig. 4, the many-core processor requires a large on-chip cache and a large memory bandwidth unless embarrassingly parallel computation algorithms can be simplified. To mitigate the negative effect of the many-core processor, the processor chip has to compute single submatrix by multiple processor cores. However, it requires a fast synchronization mechanism, because software-based synchronization is too slow for fine-grained multi-threading.

Table 1 shows the requirement of the number of arithmetic unit when each core runs at 2.5 GHz. To achieve a close-to-peak performance on the future processor chip, each core needs to execute enough instructions on all the arithmetic units of the core. On a 32-core processor, each core has to employ 64 fused multiply-add arithmetic operations (FMA), while modern commodity processors support only around 4 FMAs. To utilize such many arithmetic units, some SIMD features will be required because of the ILP wall. As shown on Table 1, longer vector length helps to reduce the number of the cores in the system. However, it reduces the efficiency

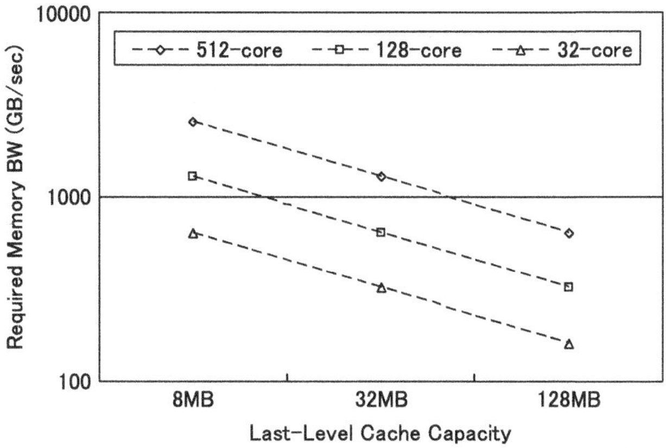

Fig. 4 Relationship between the core count and the bandwidth requirement. Typically, each core computes different part of the matrix C, but it reduces the amount of the on-chip memory as the core count increases. Unfortunately, the core count and the memory bandwidth will decrease in the future processors. Therefore, the many-core processor chip has to employ large on-chip memory and reduces the overall performance if the processor does not employ fast synchronization scheme on the chip

Table 1 Core count and SIMD width relationships on the 10 TFLOPS processor chip

	32-core	128-core	512-core
SIMD width (FMA counts/core)	64	16	4

for the other application that uses fine-grained data-level parallelism such as graph analysis.

As shown in this section, to minimize the available memory bandwidth, not many-core approach, which we call "powerful-core" approach, is better for several points, but it requires very high data-parallelism cores that handle very long vector data. If multiple cores cooperate to compute single submatrix effectively, the bandwidth requirement is reduced even on the many-core chip. However, such a fast synchronization mechanism includes technical challenges for modern processors.

5 Architecture Consideration for Exascale Supercomputing

As shown in the previous section, to minimize the memory bandwidth requirement, the powerful-core approach is often better than many-core approach, but the design of such powerful-core is very difficult. In this section, we propose the DGEMM-specialized Vector-SIMD processor that supports long vector operations to reduce the requirements of off-chip memory bandwidth. This processor is not similar to the conventional Vector-SIMD processor; the DGEMM-specialized

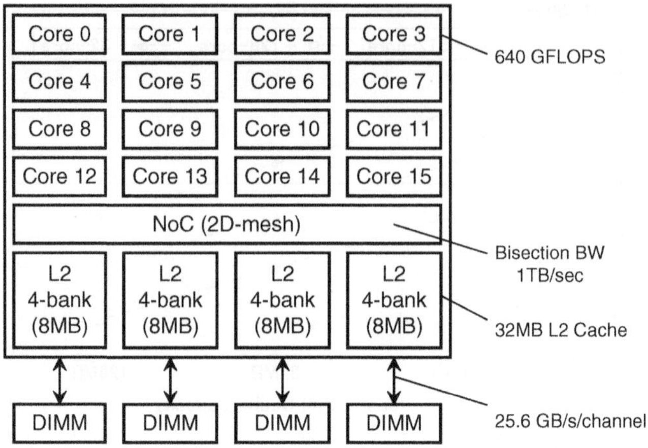

Fig. 5 Overview of CPU design for DGEMM-specialized Vector-SIMD processor. The CPU chip employs 4 DRAM channels and multiple banked cache memories. It saves the energy-consumption from conventional power-hungry vector

Vector-SIMD architecture is designed to reduce the memory bandwidth requirements for DGEMM.

Figure 5 shows the block diagram of the DGEMM-specialized Vector-SIMD processor. The processor employs 16-cores, 16-banked last-level cache, 4-channel DDR4 memory controllers, and the 2D-mesh on-chip network. The capacity of the last-level cache is 32 MB and the total off-chip memory bandwidth is 102.4 GB/s. These parameters are conservative, because the current processors have already realized these parameters. The challenge is the realization of the on-chip network having a 1 TB/s of bisection bandwidth with the 2D-mesh.

Each core of the DGEMM-specialized Vector-SIMD realizes 640 GFLOPS with 256 FMAs, as shown in Fig. 6. The supported vector length is 1024. The clock frequency of the virtual processor is 1.25 GHz. Each core also employs a 1024-word, 64-entry vector data register file and an 8-entry vector arithmetic register file. This register file organization is derived from the conventional SX architecture. The processor chip employs a fast synchronization mechanism for 16 cores.

When the DGEMM-specialized Vector-SIMD processor computes the submatrix C', the core loads C' into the vector data register as shown in Fig. 7. Each row is stored in each entry of the vector data register file. The other data are loaded on the on-chip memory. The data of matrix B are stored on the last-level cache and shared with all processor cores. The data of matrix A are stored on the scalar private cache and broadcast to each arithmetic unit. When the shape of the matrix is not square, the other blocking might be appropriate but the overall strategy is same.

As the result of such a blocking strategy, the 10-TFLOPS DGEMM-specialized Vector-SIMD processor with 32 MB on-chip memory requires only 0.006 B/F (60 GB/s) to achieve a close-to-peak performance for DGEMM.

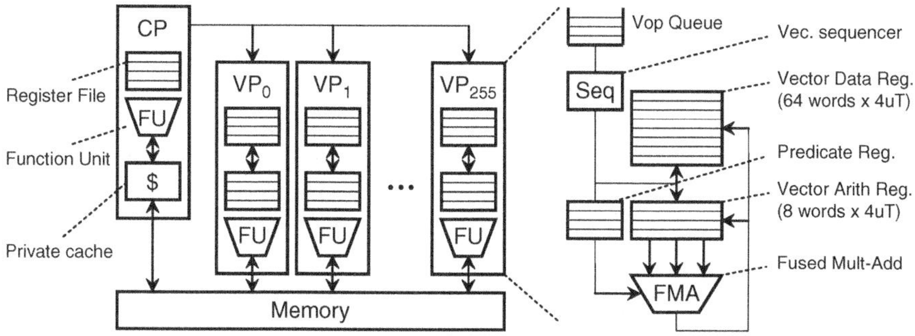

Fig. 6 Core design overview. The core employs one control processor and multiple virtual processors. Each virtual processors support four micro threads to hide the latency of the arithmetic unit

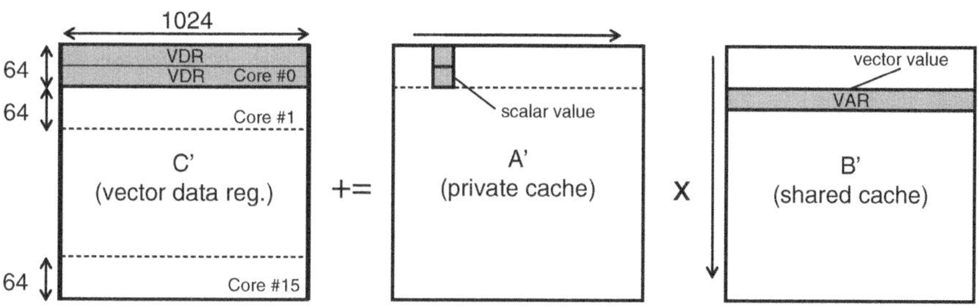

Fig. 7 Core-data mapping for DGEMM-specialized Vector-SIMD

Table 2 Energy consumption for each component

Technology	Floating-operation	Register file	Private cache	Shared cache	Main memory
15 nm	3.5 pJ/FLOP	2.29 pJ/word	1.18 pJ/word	104.8 pJ/word	9.8 pJ/bit

5.1 Comparison with Existing Architectures

In this section, we will compare the DGEMM-specialized Vector-SIMD processor with the existing architectures that are introduced in Sect. 4. We refer to the state-of-the-art DGEMM implementations to consider the energy-consumption of a future DGEMM program kernel. The parameter shown in the exascale study is used to estimate the energy consumption of the future processors. Table 2 shows the technology parameters to estimate the energy consumption of future HPC systems. Figure 8 shows the energy-consumption for each floating operations on the 15 nm process technology. As shown in the figure, the DGEMM-specialized Vector-SIMD processor can reduce the energy consumption from existing architecture design pattern.

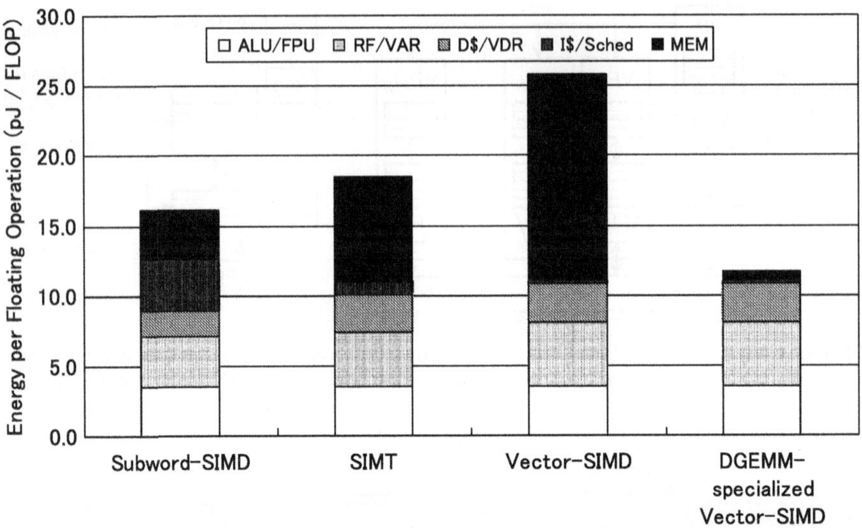

Fig. 8 Energy consumptions per FLOP of DGEMM. "ALU/FPU" is the energy consumption for arithmetic unit. "RF/VAR" is energy consumption for the register file and vector arithmetic register file. "D$/VDR" is energy consumption for the on-chip cache, the vector data register, and the on-chip network. "I$/Sched" is the energy for the instruction fetching and scheduling. "MEM" is the energy for off-chip memory access. We estimate the energy for processor chip only. Therefore, this estimation does not include any other device such as the inter-node interconnects

For further investigation, we analyze the detail of the energy consumption. Compared with Subword-SIMD processor, DGEMM-specialized Vector-SIMD processor reduces the "I$/Sched" energy for instruction scheduling by 93.8 % because Subword-SIMD processor supports short vector length (SIMD-width = 16). This SIMD width is much smaller than that of the other architecture. To reduce the energy for the instruction scheduling, the simplified pipeline and the heterogeneous processor that combines accelerators on the processor chip will be required.

The SIMT approach can reduce "I$/Sched" energy from Subword-SIMD processor, but it increases the energy for the off-chip memory access, because SIMT support many contexts on the processor chip. It shows that the many-core approach increases the memory traffic as shown in Fig. 4. To reduce the off-chip memory traffic, the many-core approach has to employ either the fast hardware synchronization scheme for hundreds of cores or the large on-chip memory. We assume these techniques will be very challenging, because existing widely-used architectures do not support such features.

The classical Vector-SIMD processor consumes much power for off-chip memory access, because it does not share submatrix B' as performed in the DGEMM-specialized Vector-SIMD processor. It increases the off-chip memory access significantly, because the programming model of the classical Vector-SIMD processor relies on wide memory bandwidth. To support the wide memory bandwidth that keeps the B/F of the current Vector-SIMD processor like the NEC SX-9, the Vector-SIMD processor has to employ more than 10 TB/s off-chip memory bandwidth on the 10-TFLOPS processor. However, it is a quite unrealistic approach, since it requires thousands of high-speed signals on a single processor chip. To overcome the limitation, some technology innovation like 3D-stacking memory technology will be required.

5.2 Discussion

As shown in the previous subsection, the existing architectures have several problems in computing DGEMM with low energy consumption, because their architectures are designed to handles the other widely various applications. To overcome the problems inherent to the existing architectures, we propose the DGEMM-specialized Vector-SIMD. It outperforms the existing architectures with several important features needed for the exascale supercomputing.

Moreover, the parameters of the DGEMM-specialized Vector-SIMD are quite realistic, considering that several current HPC systems meet these parameters. For example, GRAPE-DR supports 512 FLOP per cycle that is required to meet the core performance requirement, the SX Series supports enough large memory bandwidth that exceeds 100 GB/s, and several processors support 16 MB last-level cache. The DGEMM-specialized Vector-SIMD processor also shows that the processing of computation-intensive workloads can be accelerated through the Vector-SIMD processors.

6 Summary

The design of the computer architecture is determined from many aspects. Here underlying assumptions for such considerations sometimes lead to misunderstandings. We clarified two such misunderstandings (1) many-core approach is promising for the exascale era and (2) Vector-SIMD approach is bad for the energy-efficient computing, as shown in Sect. 4. We also introduce the DGEMM-specialized Vector-SIMD architecture, which reduces the core count to reduce the off-chip memory bandwidth to improve the energy efficiency. The DGEMM-specialized Vector-SIMD architecture outperforms existing SIMD architectures with respect to several aspects.

Towards exascale supercomputing, we have to study both applications and computer architectures. Understanding of the problem often helps to modify the existing architecture and improves the performance significantly. We are expecting that the preliminary outcomes obtained in this paper can contribute to possible approaches toward grand challenge applications on future high-performance computing systems.

References

1. Gotoblas library. http://www.tacc.utexas.edu/tacc-projects/gotoblas2/.
2. Top500 supercomputer sites. http://www.top500.org/.
3. Kazushige Goto and Robert A. van de Geijn. Anatomy of high-performance matrix multiplication. *ACM Trans. Math. Softw.*, 34(3):12:1–12:25, May 2008.
4. P. Kogge, K. Bergman, S. Borkar, D. Campbell, W. Carson, W. Dally, M. Denneau, P. Franzon, W. Harrod, K. Hill, and Others. Exascale computing study: Technology challenges in achieving exascale systems. Technical report, University of Notre Dame, CSE Dept., 2008.
5. Peter M. Kogge and Timothy J. Dysart. Using the top500 to trace and project technology and architecture trends. In *Proceedings of 2011 International Conference for High Performance Computing, Networking, Storage and Analysis*, SC '11, pages 28:1–28:11, New York, NY, USA, 2011. ACM.
6. Kuniaki Koike, Ken Fujino, Toshiyuki Fukushige, Hiroshi Daisaka, Yutaka Sugawara, Mary Inaba, Kei Hiraki, and Junnichiro Makino. Gravitational n-body simulation and lu decomposition with the multi purpose accelerator grape-dr (in japanese). *Technical Report*, 2009(26):1–11, 2009-07-28.
7. Yunsup Lee, Rimas Avizienis, Alex Bishara, Richard Xia, Derek Lockhart, Christopher Batten, and Krste Asanović. Exploring the tradeoffs between programmability and efficiency in data-parallel accelerators. In *Proceedings of the 38th annual international symposium on Computer architecture*, ISCA '11, pages 129–140, New York, NY, USA, 2011. ACM.
8. Naoya Maruyama, Masaaki Kondo, Yasuo Ishii, Akihiro Nomura, Hiroyuki Takizawa, Takahiro Katagiri, Reiji Suda, and Yutaka Ishikawa. Technical Roadmap for Exascale Supercomputing. Technical report, SDHPC, 2008.
9. Takashi Soga, Akihiro Musa, Youichi Shimomura, Ryusuke Egawa, Ken'ichi Itakura, Hiroyuki Takizawa, Koki Okabe, and Hiroaki Kobayashi. Performance evaluation of nec sx-9 using real science and engineering applications. In *Proceedings of the Conference on High Performance Computing Networking, Storage and Analysis*, SC '09, pages 28:1–28:12, New York, NY, USA, 2009. ACM.
10. Guangming Tan, Linchuan Li, Sean Triechle, Everett Phillips, Yungang Bao, and Ninghui Sun. Fast implementation of dgemm on fermi gpu. In *Proceedings of 2011 International Conference for High Performance Computing, Networking, Storage and Analysis*, SC '11, pages 35:1–35:11, New York, NY, USA, 2011. ACM.
11. David W. Wall. Limits of instruction-level parallelism. In *Proceedings of the fourth international conference on Architectural support for programming languages and operating systems*, ASPLOS-IV, pages 176–188, New York, NY, USA, 1991. ACM.
12. Wm. A. Wulf and Sally A. McKee. Hitting the memory wall: implications of the obvious. *SIGARCH Comput. Archit. News*, 23(1):20–24, March 1995.

Part II
Techniques and Tools for New-Generation Computing Systems

HPC Refactoring with Hierarchical Abstractions to Help Software Evolution

Hiroyuki Takizawa, Ryusuke Egawa, Daisuke Takahashi, and Reiji Suda

Abstract This article briefly introduces the concept of our new research project, JST CREST "An Evolutionary Approach to Construction of a Software Development Environment for Massively-Parallel Computing Systems." Since high-performance computing system architectures are going to change drastically, existing application programs will need to evolve for adapting to the new-generation systems. Motivated by this, our project will explore an effective methodology to support the programming for software evolution of valuable existing applications, and also develop a programming framework to bridge the gap between system generations and thereby to encourage migration of existing applications to the new systems. The programming framework will provide abstractions of complicated system configurations at multiple levels, and refactoring tools to help evolving applications to use the abstractions.

H. Takizawa (✉)
Graduate School of Information Sciences, Tohoku University/JST CREST, 6-6-01 Aramaki-aza-aoba, Aoba, Sendai 980-8579, Japan
e-mail: tacky@isc.tohoku.ac.jp

R. Egawa
Cyberscience Center, Tohoku University/JST CREST, 6-3 Aramaki-aza-aoba, Aoba, Sendai 980-8578, Japan
e-mail: egawa@isc.tohoku.ac.jp

D. Takahashi
Faculty of Engineering, Information and Systems, University of Tsukuba/JST CREST, 1-1-1 Tennodai, Tsukuba, Ibaraki 305-8573, Japan
e-mail: daisuke@cs.tsukuba.ac.jp

R. Suda
Graduate School of Information Science and Technology, The University of Tokyo/JST CREST, 7-3-1 Hongo, Bunkyo, Tokyo 113-8656, Japan
e-mail: reiji@is.s.u-tokyo.ac.jp

M. Resch et al. (eds.), *Sustained Simulation Performance 2012*,
DOI 10.1007/978-3-642-32454-3_3,

1 Instruction

In conventional HPC software development, the top priority is always given to performance. Lower-level programming may allow an application program to achieve a higher performance by thoroughly specializing the application code for a particular target system. However, low-level programming forces an application programmer to significantly modify the code whenever the target system changes to a new one. As a result, it is difficult to evolve existing computational science applications so as to adapt to future-generation systems, which will be massively-parallel and heterogeneous. Motivated by this, we have started a research project named "An Evolutionary Approach to Construction of a Software Development Environment for Massively-Parallel Computing Systems," which aims to support HPC software evolution adapting to system changes.

The goal of this 5.5-year project supported by JST CREST is to appropriately abstract the increasing hardware complexity of massively parallel heterogeneous computing systems, and thereby to enable computational science applications to adapt easily to new systems. This project emphasizes incremental development of existing applications and continuous software development. Therefore, the project will develop abstraction technologies so as to hide the gap between current and future systems as much as possible.

For supporting software evolution, various abstraction technologies are needed. Since we already have a huge number of valuable applications and it is impossible to completely rewrite their codes, we need to incrementally evolve them based on incremental improvement of existing programming models such as MPI and OpenMP. On the other hand, high-level abstraction is a very powerful tool to facilitate software evolution because it can hide the implementation details that are likely to be system-specific and hence major impediments to software evolution. Therefore, we will develop hierarchical abstraction layers by the following three approaches. One approach is to provide evolutionary programming models and their programming interface for massively parallel heterogeneous systems. Another is to develop numerical libraries as one high-level abstraction layer to achieve a high performance without considering the underlying system hardware. The other is to use domain-specific knowledge to build another high-level abstraction layer in order to ease application development in computational science.

We will also design a new concept of *HPC refactoring* to migrate existing application programs to new ones, which use the above hierarchical abstraction layers. Many research projects have proposed high-level descriptions of computational science applications to realize the automatic/semi-automatic translation from high-level codes to optimized low-level codes. On the other hand, software evolution in this project assumes that low-level codes already exist. The existing codes are optimized usually at a low level for current systems, not for future systems. Therefore, we first need to help migrating the existing codes to high-level ones, and then the high-level ones will be used for future maintenance and evolution while keeping their performances. The migration support, HPC refactoring, is one of the

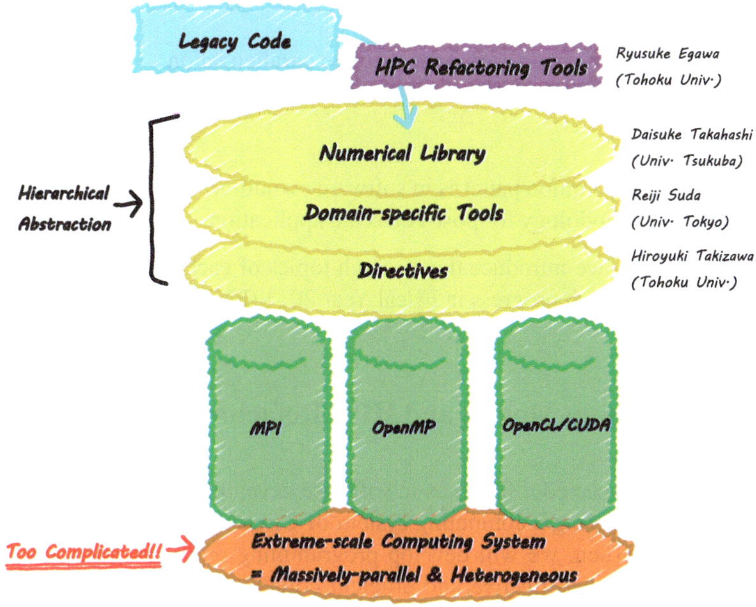

Fig. 1 Overview of the research project. As the hardware configuration of post Petascale computing systems is too complicated, this project will develop hierarchical abstraction layers to facilitate software development and future maintenance. In addition, we will establish a new concept of "HPC refactoring" for smooth migration of existing applications to the abstracted programming environment

most important features characterizing this project. We will integrate the techniques developed in this project into a programming framework, called *Xevoler*.

Since October 2011, we have started developing the above abstraction layers, and also designing the initial version of HPC refactoring catalog, which is the guideline of software evolution under the assumption of using the abstraction layers. In addition, we had a kick-off meeting and created a wiki page for project members as the infrastructure for our collaborative work.

Figure 1 shows the overview of the research project. The project team consists of the following four groups:

- Takizawa group:
  - Programming interface for managing system heterogeneity
  - Customizable refactoring tools and directives
- Takahashi group:
  - Numerical libraries to fully exploit the system performance
  - Fault tolerance and mixed-precision computation

- Suda group:
  - Domain-specific knowledge for extreme parallelization
  - Algorithm/data-structure translation for strong scaling
- Egawa group
  - Cataloging common patterns in software evolution
  - Design methodology for post-Petascale applications

In the followings, we introduce the research topics of each group, and then briefly describe their research progress in Fiscal Year 2011 (FY2011).

2 Programming Models and HPC Refactoring Tools

We discuss the expected difficulties in software development for future computing systems by considering a computing system of CPUs and GPUs as a prototype of future systems. Then, we will develop programming interfaces such as compiler directives to describe effective and efficient collaboration among many different processors. Programming models will also be designed so as to reduce the number of system-dependent parameters decided by programmers, and hence improve the code and performance portabilities of GPU computing applications.

In FY2011, we discussed the concept and future direction of this project with many researchers [1, 2]. We also developed a data dependency analysis tool [3] and a performance analysis tool [4] to help code manipulation by programmers for HPC refactoring. In addition, we developed and evaluated the mechanisms for improving the system dependability [5] and for the cache locality [6]. Those mechanisms will be key technologies for effective use of massively parallel heterogeneous systems.

3 Numerical Libraries for Heterogeneous Computing Systems

In this project, we will develop libraries of Fast Fourier Transform (FFT), Algebraic Multi-Grid (AMG), and mixed-precision basic linear algebra subprograms (BLAS). Although many large-scale applications in computational science internally use numerical libraries, most of the existing libraries are not designed considering future mainstream systems that are massively-parallel and heterogeneous. Thus, it is necessary to develop numerical libraries that can exploit the potential of massively-parallel heterogeneous systems such as large-scale GPU clusters.

In FY2011, we explored an effective implementation scheme of FFT library and prototyped a library for preliminary evaluation on a multi-core cluster system. We also considered the basic design of AMG library for GPU systems [7]. In addition, we prototyped a triple-precision BLAS library and evaluated the performance [8].

4 Use of Domain-Specific Knowledge

We explore software development methodology for parallel applications in computational science from the following two viewpoints. One is focusing on parallelization methods, and the other is on numerical calculation methods.

In FY2011, from the former viewpoint, we have developed a method to reduce collective communications in the conjugate gradient (CG) method [9]. In a standard CG method, collective communications are required twice in one iteration. However, the proposed method called the k-skip CG method needs only one collective communication in $k + 1$ iterations. Although the proposed method needs more computation and is less stable than the standard CG method, its computational complexity is less than the methods in the related work. In addition, we proposed three techniques to reduce the branch divergence, considering the importance of exploiting SIMD parallelism in future systems due to the power efficiency [10]. Those techniques will be applied to the application programs developed by Takahashi group.

From the latter viewpoint, we proposed a method to minimize the number of trials required for a Monte Carlo simulation of optimizing design parameters [11]. We also proposed a new high-order difference formula of fractional calculus [12], which is often used in the field of engineering but whose numerical method is not established yet. Moreover, we analyzed the error of QR update algorithm that can quickly solve linear least-squares problems but accumulates the errors. Then, we proposed a method to restart the update algorithm when the accumulated error exceeds a certain threshold. To explore the application design methodology in the massively-parallel heterogeneous computing era, we started analyzing important application programs in nano-science and bio-science.

In addition, for developing refactoring tools, we surveyed existing technologies in software engineering, programming models, language processing systems, and integrated development environment.

5 Design of HPC Refactoring

We are designing an HPC refactoring catalog by porting the existing applications to various platforms whose successors will potentially become the building blocks of future systems [13]. In FY2011, we have analyzed real application codes used in Tohoku University Cyberscience Center [14], surveyed code maintenance technologies in HPC software development [13], and discussed the format of HPC refactoring catalog. In those activities, we gathered the contents that should be described in the initial version of HPC refactoring catalog.

We also interviewed application programmers to collect opinions that help design of a practical HPC refactoring catalog. Furthermore, by optimizing and parallelizing existing applications, we developed optimization techniques to efficiently use the performance of parallel heterogeneous systems.

6 Conclusions

In this article, we have introduced our new research project for adapting existing applications to new-generation computing systems. In this project, we are developing various abstraction techniques to hide the hardware complexity, and also designing HPC refactoring to help migrating existing application programs to the abstracted programming environment. We will integrate these technologies into a programming framework, named *Xevolver*. Using the framework, we will help evolving various computational science applications in a systematic way.

Acknowledgements The authors would like to thank Prof. Michael Resch of HLRS, Prof. Wenmei W. Hwu of UIUC, and Prof. Chisachi Kato of the University of Tokyo for their valuable comments on this project. The authors would also like to thank Prof. Hiroaki Kobayashi of Tohoku University for constructive discussions.

This work is supported by JST CREST "An Evolutionary Approach to Construction of a Software Development Environment for Massively-Parallel Computing Systems."

References

1. Takizawa, H.: "A new research project for enabling evolution of legacy code into massively-parallel heterogeneous computing applications," The 14th Teraflop Workshop, Stuttgart, Dec 5 (2012).
2. Takizawa, H.: "How can we help software evolution for post-Peta scale computing and beyond?," The 2nd AICS symposium, Kobe, Mar 2 (2012).
3. Sato, K., Komatsu, K., Takizawa, H. and Kobayashi, H.: "A Runtime Dependency Analysis Method for Task Parallelization of OpenCL Programs," IPSJ Transactions on Advanced Computing Systems(ACS), Vol.5 No.1, pp.53–67 (2011).
4. Kanda, H., Okuyama, T., Ino, F. and Hagihara, K.: "An Instrumentation Method for Analyzing Efficiency of Memory Access in CUDA Programs," IPSJ SIG Notes 2012-HPC-133(3), 1–8, Mar 26 (2012).
5. Amrizal, M.A., Sato, K., Komatsu, K., Takizawa, H. and Kobayashi, H.: "Evaluation of a Scalable Checkpointing Mechanism for Heterogeneous Computing Systems," presentation at IPSJ Tohoku Branch Workshop, Mar 2 (2012).
6. Sugimoto, Y., Ino, F. and Hagihara, K.: "Improving Cache Locality for Ray Casting with CUDA," Proc. 25th Int'l Conf. Architecture of Computing Systems Workshops, 339–350, Feb 29 (2012).
7. Takahashi, K., Fujii, A. and Tanaka, T.: "Multiple GPUs-based AMG Method," IPSJ SIG Notes 2012-HPC-133(29), 1–7, Mar 19 (2012).
8. Mukunoki, D. and Takahashi, D.: "Implementation and Evaluation of Triple Precision BLAS Subroutines on GPUs," The 13th Workshop on Parallel and Distributed Scientific and Engineering Computing (PDSEC-12), May 25 (2012).
9. Motoya, T. and Suda, R.: "k-skip Conjugate Gradient Methods: Communication Avoiding Iterative Symmetric Positive Definite Sparse Linear Solver For Large Scale Parallel Computings," IPSJ SIG Tech. Rep. 2012-HPC-133(30), Mar. 27 (2012), in Japanese.
10. Kato, S., Suda, R. and Tamada, Y.: "Optimization Techniques for Reducing Branch Divergence on GPUs," IPSJ SIG Tech. Rep. 2012-HPC-134(5), Jun. 1 (2012), in Japanese.
11. Suda, R. and Nittoor, V.S.: "Efficient Monte Carlo Optimization with ATMathCoreLib," IPSJ SIG Tech. Rep. 2012-HPC-133(21), Mar. 27 (2012).

12. Takeuchi, Y. and Suda, R.: "New numerical computation formula and error analysis of some existing formulae in fractional derivatives and integrals," The 5th IFAC Symposium on Fractional Differentiation and its Applications (FDA'12), Keynote, May 15 (2012).
13. Egawa, R.: "Designing a Refactoring Catalog for HPC," The 15th Workshop on Sustained Simulation Performance, Sendai, Mar 23 (2012).
14. Komatsu, K., Soga, T., Egawa, R., Takizawa, H., Kobayashi, H., Takahashi, H., Sasaki, D. and Nakahashi, K.: "Performance Evaluation of BCM on Various Supercomputing Systems," In 24th International Conference on Parallel Computational Fluid Dynamics, pages 11–12 (2012).

Exploring a Design Space of 3-D Stacked Vector Processors

Ryusuke Egawa, Jubee Tada, and Hiroaki Kobayashi

Abstract Three dimensional (3-D) technologies have come under the spotlight to overcome limitations of conventional two dimensional (2-D) microprocessor implementations. However, the effect of 3-D integrations with vertical interconnects in future vector processors design is not well discussed yet. In this paper, aiming at exploring the design space of future vector processors, fine and coarse grain 3-D integrations that aggressively employ vertical interconnects are designed and evaluated.

1 Introduction

Modern vector processors play important roles in high performance computing due to the significant advantages over commodity-based scalar processors for memory-intensive scientific and engineering applications. However, vector processors still keep a single core architecture, though chip multiprocessors (CMPs) have become the mainstream in recent processor architectures. Twelve-cores CMPs are already in the commercial market, and an 80-cores CMP is prototyped by Intel to overcome power and performance limitations of single core architectures [30]. On the other hand, CMP-based vector processors have not been found as real products. However, the CMP architecture is also promising for vector processor design, because recent scientific and engineering applications running on a vector supercomputer are well

R. Egawa (✉) · H. Kobayashi
Cyberscience Center, Tohoku University/JST CREST, 6-3 Aramaki-aza-aoba, Aoba, Sendai 980-8578, Japan
e-mail: egawa@isc.tohoku.ac.jp; koba@isc.tohoku.ac.jp

J. Tada
Graduate School of Science and Engineering, Yamagata University, 4-3-16 Jonan, Yonezawa 992-8510, Japan
e-mail: jubee@yz.yamagata-u.ac.jp

M. Resch et al. (eds.), *Sustained Simulation Performance 2012*,
DOI 10.1007/978-3-642-32454-3_4, © Springer-Verlag Berlin Heidelberg 2013

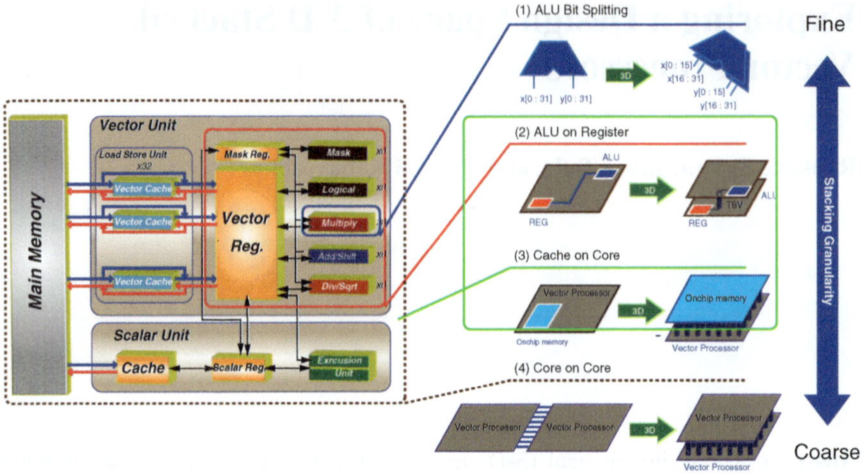

Fig. 1 Stacking granularities

parallelized by vector compilers and/or OpenMP. Under this situation, a Chip Multi Vector Processor (CMVP) architecture has been proposed by Musa et al., and its potential has been clarified [22].

Essentially, vector processors need a plenty of hardware resources compared to scalar processors. This is because a vector processor installs many vector pipelines, large vector registers and vast I/O logics to simultaneously process a huge amount of data provided at a high-memory bandwidth [9]. In addition, CMVP has a large shared on-chip memory named a *vector cache*. Therefore, even if technology scaling were to advance as ever, it would be difficult to implement CMVP with many vector cores by the conventional 2-D implementation technology due to the area limitation.

Recently, 3-D integration technologies have come under the spotlight to overcome the limitations of conventional 2-D microprocessor implementations. 3-D integration technologies are expected to greatly increase transistor density while providing faster on-chip communication. Three dimension integration technologies are not a brand new, and various 3-D integration technologies have been explored, such as micro-bump, wire bonding and through via vertical interconnect etc [16]. Among these various technologies, a vertical interconnect with though-silicon-via (TSV) is assumed as the most promising one to expand the design space of future high-performance and low-power microprocessors [13]. Thus computer architects and circuit designers are re-attracted to 3-D integration technologies by an appearance of TSVs with high feasibility.

To introduce 3-D Die stacking technologies into the future vector processors design, several granularities can be considered as shown in Fig. 1. Aiming at clarifying the potential of 3-D integration technologies, this paper examines fine grain and coarse grain 3-D designs. The fine grain denotes logic level 3-D integrations such as 3-D stacked arithmetic units designs. On the other hand, the coarse grain stacks more large chunks such as cores or memories. As examples of the fine grain 3-D integrations, we focus on the floating point arithmetic units, which are the key components of vector processors. On the other hand, CMVP architecture is selected as a target of the coarse grain 3-D integrations. To realize the 3-D integration of CMVP, CMVP is modified to exploit the potential of 3-D integration technologies. In this paper, through these designs and early evaluations, a design space of future vector processors is explored.

This paper is organized as follows. Section 2 briefly outlines the basis of 3-D integration technologies and related works. Then 3-D integrated arithmetic units are designed and evaluated in Sect. 3. In Sect. 4, a 3-D stacked CMVP is introduced, and its early performance evaluations are carried out. Section 5 concludes this work.

2 3-D Die Stacking Technologies

2.1 Die Stacking with TSVs

3-D integration is emerging fabrication technology that vertically stacks multiple integrated chips. The benefits include an increase in device density, the flexibility of routing, and the ability to integrate disparate technologies [4, 16]. Although there are a lot of technologies such as wire bonding to realize a vertical stacking of two or more integrated circuits, this study focuses on 3-D integrated circuits with vertical interconnects due to its short delay with high density. So far, many researches have reported processing technologies for thin and long TSVs with high feasibility [1, 3, 8, 12, 18]. In addition, some researchers have clarified the potential of TSVs by analyzing its electrical characteristics [11, 23].

Two topologies can be conceived to bound two silicon dies: *face-to-face* and *face-to-back*, where face is the side with the metal layer and back is the side with the silicon substrate as shown in Fig. 2. In the face to face bonding, die to die (D2D) vias are processed and deposited on top of metal layers as the conventional metal etching technologies. Although face to face bonding can provide higher D2D via density and lower area overhead than face to back, it can just allow to stack two active silicon layers. On the other hand, face to back bonding can stack any number of multiple active silicon layers by TSVs that go through silicon bulk with lower via density. As noted in earlier studies, the dimension for TSVs vary from 2 to 5 μm in recent real implementations and pitch of 3-D via only requires a modest overhead. More details description about processing techniques of TSVs are described in [12, 18]. Thus, to realize aggressive stacking of multiple layers, face to back seems preferable. In this paper, the following circuits and processors are designed using two to eight silicon layers, henceforth, we focus on the face to back wafer bonding technique.

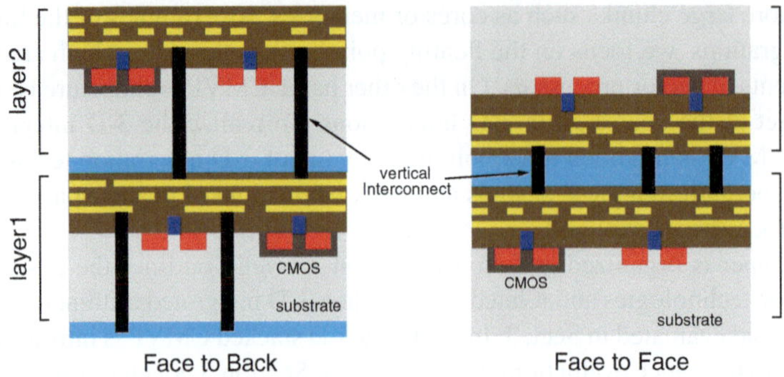

Fig. 2 Face to back vs. Face to face

2.2 Related Work

According to the trend of recent microprocessor designs, several researches have combined 3-D memory stacking and CMPs to supply data to a chip with a massive number of cores at an enough bandwidth [10, 15]. Black et al. [2] have explored the memory bandwidth benefits using Intel Core2 Duo processor. By stacking an additional cache memory layer, the on-chip cache capacity is increased, and the performance is improved by reducing off-chip memory accesses. Loh et al. [17] have discussed DRAM stacking on CMPs. This paper has tried to fully take advantages of the benefit of 3-D integration technologies with TSVs, memory organizations are optimized for many purposes such as the main memory or the last level caches. In addition, stacking layers of non-volatile cache memory such a Phase Change Random Access Memory (PRAM) [31] and a Magnetic Random Access Memory (MRAM) [5] are also studied to mitigate the effects of the processor-memory speed gap with low power consumption. However, these researches are carried out based on multi-core or many-core scalar microprocessors. In this paper, we focus on the chip multi-vector processor, which cannot be implemented by conventional 2-D technologies with a high memory bandwidth. To realize a high sustained performance by improving the memory bandwidth, the 3-D integration technology is suitable for the vector architecture. Thus this paper targets on designing a 3-D stacked CMVP as an example of a coarse grain die stacking implementation.

On the other hand, there are few studies that try to clarify the effects of fine grain 3-D integrations. Mayage et al. [19] have designed a 3-D carry look ahead adder with face to back implementation. Puttaswamy and Loh also have explored the effects of employing the vertical interconnect in combinational logic design [24, 25]. They have designed and evaluated three kinds of parallel prefix adder and barrel shifter by the face to face implementation. In [29], they have designed and evaluated the performance of several arithmetic and control units based on face to back implementation, and reported that 3-D integrated circuits can improve their power efficiency and performance in the future CMOS and 3-D integration technologies. Due to the small number of researches in this field, these explorations are limited to small-scale arithmetic units, though 3-D integration seems effective in a large-scale circuit design. Hence, this paper selects floating point arithmetic units, which are the most largest and important combinational logic circuits in the recent high-performance microprocessors.

3 3-D Stacked Arithmetic Units

In recent microprocessors, floating-point arithmetic units play important role to achieve a high computational performance. Especially, vector processors and GPUs such as processors with SIMD/Vector instructions install a lot of floating-point arithmetic units to achieve a high computational performance. Thus, in this section, floating point arithmetic units are design in a 3-D fashion and evaluated. Through these approaches, the effectiveness and potential of 3-D integration technologies in floating point unit design are clarified. To design 3-D stacked arithmetic units, first, an arithmetic unit is partitioned into some sub-circuits, based on a circuit partitioning strategy. Next, each sub-circuit is placed on one layer, and data-transfers between sub-circuits are done through TSVs. TSVs require Keep-out zone to avoid electrical effect caused by TSVs to transistors.

Therefore, to relax this effect, it is assumed that the area of TSVs and the area of sub-circuits are separated, and TSVs are placed around the sub-circuits as shown in Fig. 3. More detailed design flow can be confirmed in [6].

Based on the discussions in our previous work, there are two requirements for a partitioning strategy for 3-D integration. First, sub-circuits should have almost same size for minimize the footprint, and the critical-path delay should not be enlarged. To fulfill these requirements, we use a circuit partitioning strategy proposed in [26]. The concept of this strategy is shown in Fig. 4, and the features of the strategy are as follows:

- Gates on the critical-path are packed in single layer.
- The area of each layer is equalized as much as possible.
- Small components should be packed together on one layer.

If the critical-path in a circuit is divided into some sub-circuits, TSVs are inserted on the critical-path, and it will increase the maximum delay of a circuit due to its large capacitance. However, it is difficult to partition small components into more small sub-circuits without dividing the critical-path. Therefore, large components should be partitioned into sub-circuits without dividing its critical-path, and small components should be packed together into one layer. To keep the footprint of 3-D stacked VLSIs small, the size of each layer should be equalized. If one sub-circuit is larger than other sub-circuits, it enlarges the footprint of 3-D stacked VLSI. This equalization is controlled and achieved by partitioning large components.

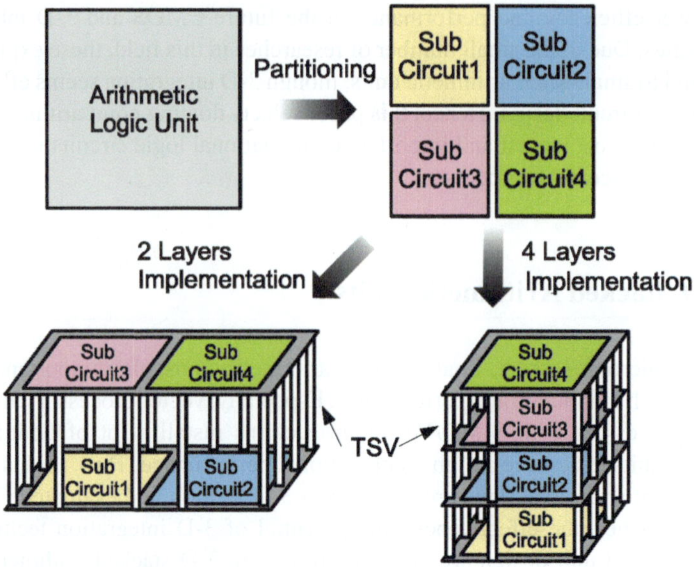

Fig. 3 Design flow of 3-D stacked arithmetic units

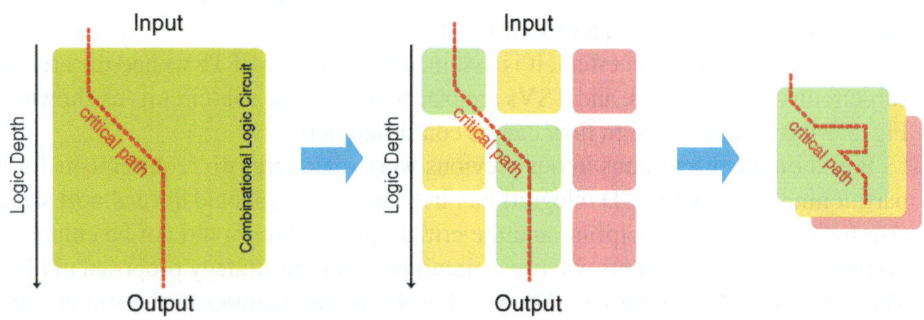

Fig. 4 Circuit partitioning for 3D die stacking

Figure 5 shows a circuit partitioning for a floating-point multiplier. The layer 1 includes the lower-bits part of a booth-encoder and a wallace-tree, and other components of the floating-point multiplier. Other layers have the rest parts of the booth-encoder and the wallace-tree. This is because a significant multiplier, which consists of the booth-encoder, the wallace-tree and the final adder occupies large part of the circuit area. Thus, the booth-encoder and the wallace-tree are partitioned into some sub-circuits. Since the final adder is not partitioned due to avoid dividing the critical-path, the adder is implemented into a single layer.

To evaluate the effects of 3-D Die stacking technologies in arithmetic units designs, single and double precision 3-D stacked floating multipliers are designed using the 180 nm CMOS technology with TSVs. The parameters of TSVs are

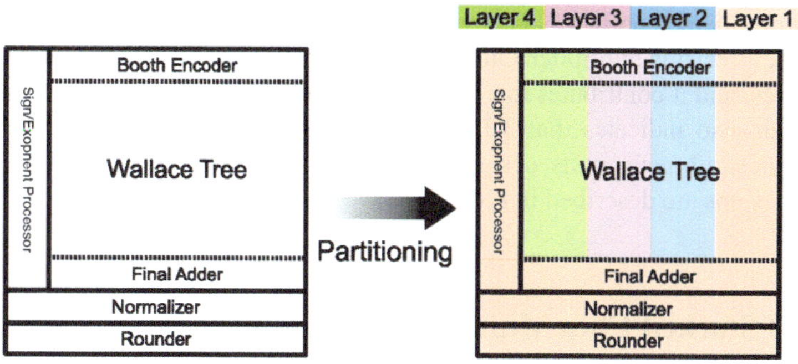

Fig. 5 Design of 3-D stacked floating multiplier (4 layers)

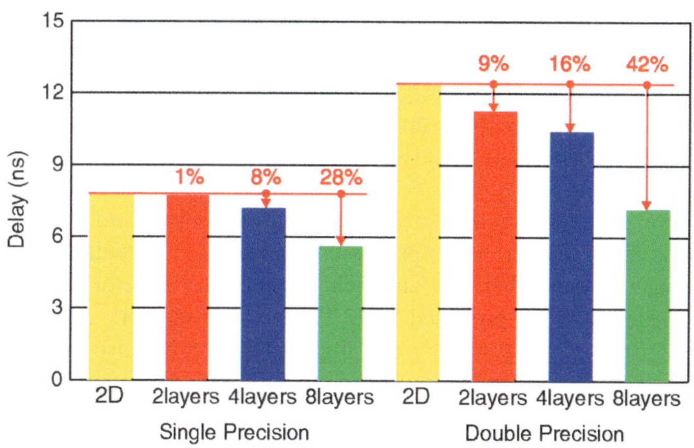

Fig. 6 Effects of 3-D die stacking

assumed based on the ITRS reports, and the resistance and capacitance of the TSVs are set to 7 mΩ and 27 fF, respectively. The number of layers is varied from two to eight. Figure 6 shows the maximum delay of 3-D stacked floating point multipliers. Compared to 2-D implementations, 3-D implementations achieve significant delay reductions. Both in the cases of single and double precision designs, reductions in the maximum delay are boosted as the number of layers increases. This is mainly because the size of layers becomes small as the number of layers increases. This also eliminates the extremely long wires and shortens the average wire length in each design, hence 3-D integrations achieve up to 42 % delay reduction. From these results, we can confirm that 3-D integration technologies with TSVs have enough potential to reduce the maximum delay of floating point arithmetic units by selecting appropriate circuit partitioning strategies.

In addition the maximum delay reduction in the double precision multipliers are larger than those of single-precision multipliers. In the double-precision floating point multipliers, long wires occupy the many parts of total wire length compared to the single-precision multiplier. Our 3-D implementation reduces the number of these long wires, and it contributes to a reduction in the maximum path delay. Therefore, this result also indicates that 3-D integration technologies are more effective in large-scale arithmetic units designs. More details of the fine grain 3-D stacked circuit designs are described in [26].

4 3-D Stacked Chip Multi Vector Processors

4.1 An Overview of 3-D Stacked CMVP

Recently, it is getting harder to further improve the performance and energy efficiency of vector processors due to several limitations such as the die area and the number of I/O pins on a chip. The most severe problem is the decrease in the ratio of memory bandwidth to the floating-point operation rate (Bytes/Flop, B/F). A high B/F ratio is essential to achieve a high computational efficiency, i.e., efficient use of the computing power. If the B/F ratio of a vector processor decreases to be as the same level as that of a scalar one, the vector processor will no longer be able to keep its superiority over the scalar processor in terms of the sustained performance. To compensate for a low B/F ratio, an on-chip memory for a vector processor named a vector cache has been proposed [21]. The on-chip memory can provide data to vector registers of the processor at a high bandwidth because the data transfer does not need I/O pins. Hence, the B/F ratio of a vector processor is improved by storing reusable data in the on-chip memory to achieve high sustained performance. In addition, an on-chip memory is expected to decrease the number of off-chip main memory accesses, resulting in decreasing the energy consumption in the processor I/O, memory network and off chip memory components. Therefore, the vector cache is also introduced into the 3-D stacked CMVP. The vector cache is not private to each processor core but shared by multiple cores because scientific simulations such as difference schemes often have a high locality of memory reference among threads.

Figure 7 shows the basic structure of the 3-D stacked CMVP. The 3-D stacked CMVP is composed of three kinds of layers; I/O layer, core layer, vector cache layer. The I/O layer contributes to keep off-chip memory bandwidth, and the core layer realizes implementation of many cores on a die. The vector cache layer works for increasing the capacity of on-chip memory to compensate for insufficient memory bandwidth of each core.

The core layer includes two vector cores, and each vector core is designed based on the NEC SX-8 processor. The vector core has four parallel vector pipe sets, each of which contains six types of vector arithmetic pipes (Mask, Logical, Add/Shift, Multiply, Divide, Square root), and 144 KB vector registers as shown in Fig. 8.

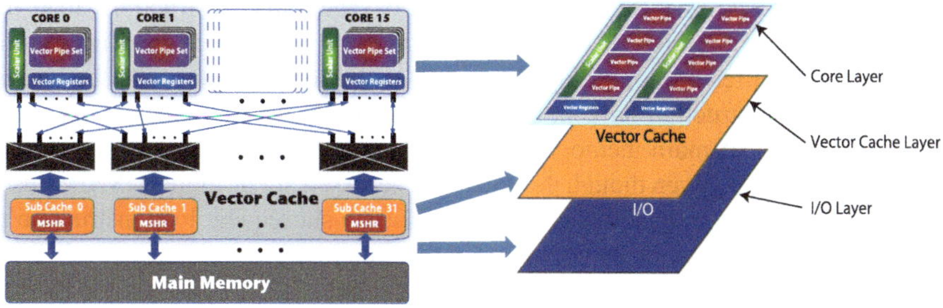

Fig. 7 Basic structure of the 3-D stacked CMVP

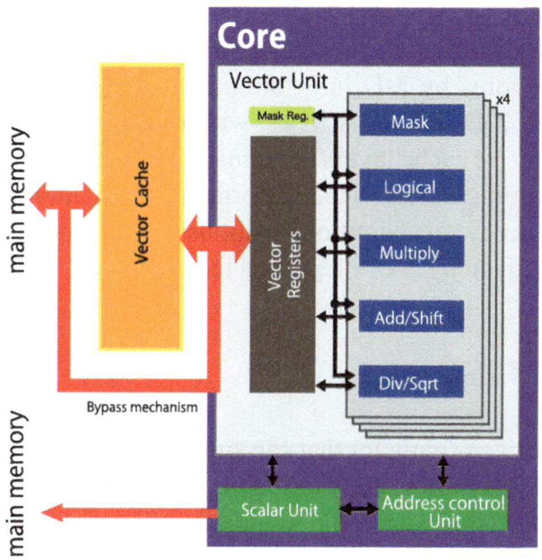

Fig. 8 Block diagram of a core

On the I/O layer, only SERializer/DESerializer(Ser/Des logic) is implemented. As shown in [9], the occupancies of Ser/Des logic in the NEC vector processors SX-8 and SX-9 are quite high. Thus the 3-D stacked CMVP introduces independent I/O layers to keep a high memory bandwidth.

The vector cache layers are put between the I/O layers and the core layers. The vector cache layers consist of 32 2-way set-associative sub-caches with miss status handling registers (MSHR) [14]. A vector load/store instruction of the vector architecture is able to concurrently deal with up to 256 floating-point data. Hence, the vector cache also needs to process up to 256 data in continuity. Furthermore, the vector cache employs a bypass mechanism between the main memory and vector register files. The bypass mechanism makes possible to supply data from both the main memory and the vector cache at the same time. Thus, the total amount of data provided to the vector register files in time is increased by the bypass mechanism. In addition, the vector cache is a non-blocking cache with MSHR. In scientific computations such as difference schemes, two vector load instructions load the memory regions that are partially overlapped. If the subsequent load instruction is issued right after the preceding instruction, however, the data to be fetched by the preceding instruction have not been cached in the vector cache yet owing to the long latency of main memory accesses. Thus, the subsequent load instruction causes cache misses even though the data to be accessed are in-flight. MSHR is used to avoid this situation, and makes it possible for the subsequent load instruction to reuse in-flight data fetched by the preceding instruction.

4.2 Performance Evaluations

In this section, to clarify the effects of increasing the off-chips memory bandwidth and the number of cores by 3-D integration technologies, we firstly evaluate the performance of the 3-D stacked CMVP without vector cache layers. Then the performance with vector cache layers is examined in terms of sustained performance and energy consumption. Based on the performance evaluation, the tradeoff between performance and energy consumption is discussed to realize effective usage of a plenty of hardware given by 3-D integration technologies.

4.2.1 Evaluation Setup

An NEC SX trace-driven simulator that can simulate the behavior of the 3-D stacked CMVP architecture at the register-transfer level is implemented. The simulator is designed based on the NEC SX vector architecture. The simulator accurately models a vector core of the SX-8 architecture; the vector unit, the scalar unit and the memory system. The simulator takes system parameters and a trace file of benchmark programs as input, and outputs instruction cycle counts. The specification of 3-D stacked CMVP is shown in Table 1. We assume that the I/O layer provides a 64 GB/s memory bandwidth with one layer, the number of cores is possible at the maximum of 16, and the maximum capacity of the vector cache is 32 MB. Since we assume that the TSV has 2 um diameter with a 30 um length [13], the access latency between cores and the vector cache reduces to 70 % of the cache access latency of conventional 2-D implementations. The energy consumed by the vector cache accesses are obtained by CACTI6.5 [20].

The evaluations are performed by using a FDTD code, which simulates the antipodal fermi antenna [27] by using SX-9 of Tohoku University. The number of grids, the vector operation ratio and vector length are $612 \times 105 \times 505$, 99.9% and 255, respectively. The benchmark programs is compiled by the NEC FORTRAN compiler, which can vectorize and parallelize the applications automatically. Then executable programs run on the SX trace generator to produce the trace files.

Table 1 Specification of 3-D stacked CMVP

Parameter	Value
Number of cores	1–16
Vector cache implementation	SRAM
Capacity of the vector cache	512–32 MB
Cache line size	8B
Cache policy	Write-through, LRU replacement
Cache associativity	2
Memory bandwidth (between cache and core)	64 GB/s/core
Off-chip Memory bandwidth	64–256 GB/s
Tr. process technology	90 nm
Number of entries of MSHR	8,192

4.2.2 Evaluation Results and Discussions

First, the performance of the 3-D stacked CMVP with the various number of cores is evaluated. As shown in the previous section, since one core layer includes two vector cores, 4-cores, 8-cores and 16-cores are implemented by two layers, 4-layers and 8-layers, respectively. The off-chip memory bandwidth is also varied from 64 to 256 GB/s by stacking the necessary number of I/O layers. In the case of using a single I/O layer, the off-chip memory bandwidths are 64 GB/s and 128 GB/s by changing the usage of silicon budgets. 64 GB/s is achieved when a half part of the I/O layer is used, and 128 GB/s is achieved by using the whole I/O layer. Doubling I/O layers realizes 256 GB/s. The memory bandwidth per core is decreased as the number of cores increases, thus 4 B/F rate per core is achieved in the cases of 2 cores with 128 GB/s of the off-chip memory bandwidth and 4 cores with 256 GB/s of the off-chip memory bandwidth.

Figure 9a shows the performance of the 3-D stacked CMVP when changing the off-chip memory bandwidth. The performance is normalized by the case of single core, without vector cache, with baseline off-chip memory bandwidth (64 GB/s). In this graph, yellow, blue and red bars indicate the cases of 64, 128, and 256 GB/s, respectively. From these results, we can confirm that enhancing off-chip memory bandwidth improves the sustained performance with a high-scalability. On the other hand, Fig. 9b shows the performance of the 3-D stacked CMVP with a 8 MB vector cache when changing the off-chip memory bandwidth. We can also confirm that the vector cache has a high potential to improve the performance in the all the cases. These results indicate that there are two choices to improve the performance of the 3-D stacked CMVP by enhancing off-chip.

Next, effects of enhancing off-chip memory bandwidth by introducing I/O layers and employing the vector cache are discussed. Figure 10a shows the normalized performances of 16 cores cases. The performance of using two I/O layers without the vector cache (1 B/F) and using one I/O layer with an 8 MB vector cache (0.5 B/F + Cache) are comparable. Both cases improve the effective memory

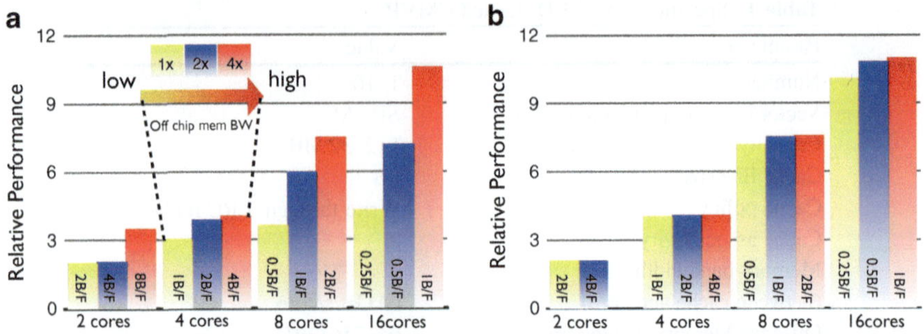

Fig. 9 Effects of enhancing off-chip memory bandwidth and vector cache. **(a)** Effects of off-chips mem. bandwidth. **(b)** Effects of 8 MB vector cache

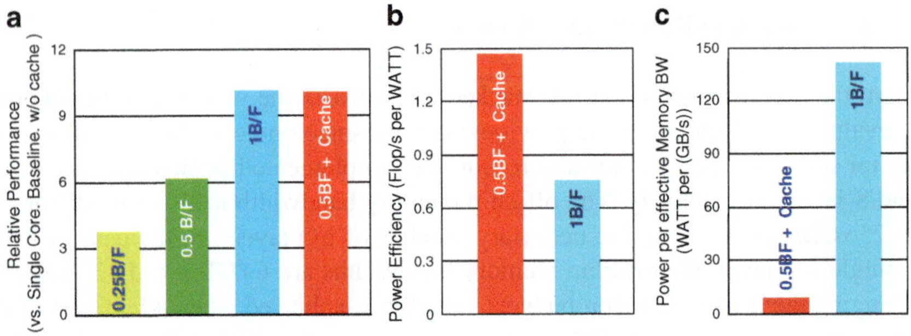

Fig. 10 Off-chip vs. On-chip memory bandwidth. **(a)** Performance. **(b)** Sustained Performance per Watt. **(c)** Watt per Effective Memory BW

bandwidth, and then improve the performance compared to the base-lined case. This result indicates that several architectural designs of the 3-D stacked CMVP can be considered. To clarify the energy efficient configurations of the 3-D stacked CMVP, the power and energy consumption of the 3-D stacked CMVP are evaluated. Figure 10b shows the power efficiency of the 3-D stacked CMVP. The power efficiency is obtained by *performance/Watt*. This result indicates that the vector cache can achieve 49 % higher power efficiency (*performance/watt*) compared to the case of increasing the off-chip memory bandwidth. In addition, the vector cache requires a 91 % smaller power consumption per effective memory bandwidth compared to the case of increasing the off-chip memory bandwidth as shown in Fig. 10c. Therefore, to realize energy efficient computing on the 3-D stacked CMVP, the vector cache is much more power-efficient. Since the power efficiency would strongly depend on applications, more variable evaluations using other benchmarks are needed to determine the best configuration in terms of the performance and the power consumption. This consideration remains as our future work.

In this paper, we set a bandwidth between the vector cache and the vector register as 64 GB/s/core. However, there are several researches designing 3-D stacked cache memories, which realize a high memory bandwidth and huge capacity using

Fig. 11 Effect of an enhanced the vector cache on the performance and energy

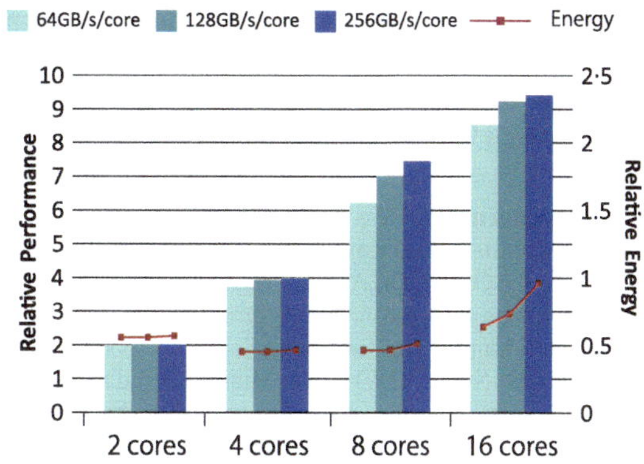

through silicon vias (TSVs) [15, 28]. Since TSVs have small RC delay compared to conventional 2-D wire and recent TSVs process technologies allow to implement many TSVs on a chip [8], 3-D stacked cache memories have potential to realize higher memory bandwidth compared to 2-D implementation [7]. Thus, we can assume the future 3-D stacked vector cache have a high memory bandwidth of 128 GB/s/core and 256 GB/s/core.

Figure 11 shows the performance and energy consumption of the 3-D stacked CMVP with a 8 MB enhanced vector cache. In this evaluation, the FDTD code is also used as a benchmark, and every value are normalized by that of a single core without the vector cache. From these results, we can confirm that enhancing the performance of the vector cache is quite effective to improve energy efficiencies of the 3-D stacked CMVP. More detail design of the 3-D vector cache, which exploits the potential of 3-D die stacking technologies should be considered as our future work.

5 Conclusions

To clarify the potential of 3-D Die stacking in the future vector processors design, fine and coarse grain 3-D integrations are examined. As a fine grain 3-D integration, 3-D stacked floating point multipliers are designed and evaluated. By partitioning the floating point multipliers appropriately, 3-D integration technologies can significantly reduce the maximum delay of the floating point multipliers. On the other hand, as a coarse grain 3-D integration, the 3-D stacked CMVP is introduced and evaluated. The effects of the vector cache and enhancing off-chip memory bandwidth by I/O layers are evaluated from the viewpoint of the performance and the energy consumption. Evaluation results show that the vector cache can effectively decrease the power and energy consumption of the 3-D stacked CMVP, while achieve a high performance. From these results, we can confirm that 3-D integration technologies have enough potential to boost the performance and the energy efficiency of future vector processors.

To realize more powerful computing environments with extremely high memory bandwidth, more detailed design of the vector cache should be considered. In addition, designing of the 3-D stacked CMVP the vector cache under the fine and coarse grain 3-D integration technologies should be considered.

Acknowledgements The authors would like to thank Associate Professor Hiroyuki Takizawa, Professor Mitsumasa Koyanagi of Tohoku University, Yusuke Funaya of Hitachi, Ryu-ichi Nagaoka of BOSCH, Dr. Akihiro Musa, Jun Inasaka and Dr. Shintaro Momose of NEC for valuable discussions on this research. This research was partially supported by Grant-in-Aid for Scientific Research (Grant-in-Aid for Young Scientists (B) No. 22 700044) and (Grant-in-Aid for Scientific Research (B) No. 22300013), the Ministry of Education, Culture, Sports, Science and Technology. This research was also partially supported by Core Research for Evolutional Science and Technology (CREST), Japan Science and Technology Agency (JST).

References

1. E. Beyne. Tsv technology overview. In *Semicon Taiwan 2008 CTO Forum*, 2008.
2. B. Black, M. Annavaram, N. Brekelbaum, J. DeVale, L. Jiang, G. H. Loh, D. McCaule, P. Morrow, D. W. Nelson, D. Pantuso, P. Reed, J. Rupley, S. Shankar, J. Shen, and C. Webb. Die stacking (3d) microarchitecture. In *MICRO 39: Proceedings of the 39th Annual IEEE/ACM International Symposium on Microarchitecture*, pages 469–479, 2006.
3. S. Das, A. Fan, K.-N. Chen, C. S. Tan, N. Checka, and R. Reif. Technology, performance, and computer-aided design of three-dimensional integrated circuits. In *ISPD '04: Proceedings of the 2004 international symposium on Physical design*, pages 108–115, New York, NY, USA, 2004. ACM.
4. W. Davis, J. Wilson, S. Mick, J. Xu, H. Hua, C. Mineo, A. Sule, M. Steer, and P. Franzon. Demystifying 3d ics: the pros and cons of going vertical. *Design & Test of Computers, IEEE*, 22(6):498–510, Nov.-Dec. 2005.
5. X. Dong, X. Wu, G. Sun, Y. Xie, H. Li, and Y. Chen. Circuit and microarchitecture evaluation of 3D stacking magnetic RAM (MRAM) as a universal memory replacement. In *DAC '08: Proceedings of the 45th annual Design Automation Conference*, pages 554–559, New York, NY, USA, 2008. ACM.
6. R. Egawa, Y. Funaya, R. Nagaoka, Y. Endo, A. Musa, H. Takizawat, and H. Kobayashi. Effects of 3-D Stacked Vector Cache on Energy Consumption. In *2011 IEEE International 3D Systems Integration Conference (3DIC)*, pages 1–8, 2012.
7. Y. Funaya, R. Egawa, H. Takizawat, and H. Kobayashi. 3D On-Chip Memory for the Vector Architecture. In *2009 IEEE International 3D Systems Integration Conference (3DIC)*, pages 1–8, 2009.
8. S. Gupta, M. Hilbert, , S. Hong, and R. Patti. Techniques for producing 3d ics with high-density interconnect. In *Proceedings of the 21st International VLSI Multilevel Interconnection Conference*, 2004.
9. J. Inasaka and M. Kajita. Techniques for power supply noise management in the SX supercomputers. In *IEICE Tech. Report*, pages 41–46, 2008.
10. T. Kgil, A. Saidi, N. Binkert, S. Reinhardt, K. Flautner, and T. Mudge. Picoserver: Using 3d stacking technology to build energy efficient servers. *J. Emerg. Technol. Comput. Syst.*, 4(4):1–34, 2008.
11. D. Khalil, Y. Ismail, M. Khellah, T. Karnik, and V. De. Analytical model for the propagation delay of through silicon vias. In *ISQED '08: Proceedings of the 9th international symposium on Quality Electronic Design*, pages 553–556, 2008.

12. M. Koyanagi, T. Fukushima, and T. Tanaka. High-density through silicon vias for 3-d lsis. *Proceedings of the IEEE*, 97(1):49–59, Jan. 2009.
13. M. Koyanagi, T. Nakamura, Y. Yamada, H. Kikuchi, T. Fukushima, T. Tanaka, and H. Kurino. Three-dimensional integration technology based on wafer bonding with vertical buried interconnections. *IEEE Trans. Electron Devices*, 53(11):2799–2808, 2006.
14. D. Kroft. Lockup-Free Instruction Fetch/Prefetch Cache Organization. *ISCA*, pages 81–88, 1981.
15. G. H. Loh. 3d-stacked memory architectures for multi-core processors. In *ISCA '08: Proceedings of the 35th International Symposium on Computer Architecture*, pages 453–464, 2008.
16. G. H. Loh, Y. Xie, and B. Black. Processor Design in 3D Die-Stacking Technologies. *IEEE Micro*, 27(3):31–48, 2007.
17. G. H. Loh, Y. Xie, and B. Black. Processor Design in 3D Die-Stacking Technologies. *Micro, IEEE*, 27(3):31–48, may. 2007.
18. P. Marchal, B. Bougard, G. Katti, M. Stucchi, W. Dehaene, A. Papanikolaou, D. Verkest, B. Swinnen, and E. Beyne. 3-d technology assessment: Path-finding the technology/design sweet-spot. *Proceedings of the IEEE*, 97(1):96–107, Jan. 2009.
19. J. Mayega, O. Erdogan, P. M. Belemjian, K. Zhou, J. F. McDonald, and R. P. Kraft. 3d direct vertical interconnect microprocessors test vehicle. In *GLSVLSI '03: Proceedings of the 13th ACM Great Lakes symposium on VLSI*, pages 141–146, New York, NY, USA, 2003. ACM.
20. N. Muralimanohar, R. Balasubramonian, and N. P. Jouppi. CACTI 6.5. Technical Report HPL-2009-85, HP Labs, 2009.
21. A. Musa, Y. Sato, R. Egawa, H. Takizawa, K. Okabe, and H. Kobayashi. An On-chip Cache Design for Vector Processors. In *MEDEA '07: Proceedings of the 2007 workshop on MEmory performance*, pages 17–23, New York, NY, USA, 2007. ACM.
22. A. Musa, Y. Sato, T. Soga, K. Okabe, R. Egawa, H. Takizawa, and H. Kobayashi. A shared cache for a chip multi vector processor. In *MEDEA '08: Proceedings of the 9th workshop on MEmory performance*, pages 24–29, New York, NY, USA, 2008. ACM.
23. J. S. Pak, C. Ryu, and J. Kim. Electrical characterization of trough silicon via (tsv) depending on structural and material parameters based on 3d full wave simulation. In *Electronic Materials and Packaging, 2007. EMAP 2007. International Conference on*, pages 1–6, Nov. 2007.
24. K. Puttaswamy and G. Loh. The impact of 3-dimensional integration on the design of arithmetic units. In *Proceeding of the IEEE International Symposium on Circuits and Systems (ISCAS)*, pages 4951–4954, May 2006.
25. K. Puttaswamyt and G. H. Loh. Scalability of 3d-integrated arithmetic units in high-performance microprocessors. In *DAC '07: Proceedings of the 44th annual Design Automation Conference*, pages 622–625, New York, NY, USA, 2007. ACM.
26. J. Tada, R. Egawa, K. Kawai, H. Kobayashi, and G. Goto. A Middle-Grain Circuit Partitioning Strategy for 3-D Integrated Floating-Point Multipliers. In *2011 IEEE International 3D Systems Integration Conference (3DIC)*, pages 1–8, 2012.
27. Y. Takagi, H. Sato, Y. Wagatsuma, K. Mizuno, and K. Sawaya. Study of High Gain and Broadband Antipodal Fermi Antenna with Corrugation. In *2004 International Symposium on Antennas and Propagation*, pages 69–72, 2004.
28. Y.-F. Tsai, F. Wang, Y. Xie, N. Vijaykrishnan, and M. J. Irwin. Design space exploration for 3-d cache. *IEEE Trans. Very Large Scale Integr. Syst.*, 16(4):444–455, 2008.
29. B. Vaidyanathan, W.-L. Hung, F. Wang, Y. Xie, V. Narayanan, and M. J. Irwin. Architecting microprocessor components in 3d design space. In *VLSID '07: Proceedings of the 20th International Conference on VLSI Design held jointly with 6th International Conference*, pages 103–108, Washington, DC, USA, 2007. IEEE Computer Society.
30. S. Vangal, J. Howard, G. Ruhl, S. Dighe, H. Wilson, J. Tschanz, D. Finan, P. Iyer, A. Singh, T. Jacob, S. Jain, S. Venkataraman, Y. Hoskote, and N. Borkar. An 80-Tile 1.28TFLOPS Network-on-Chip in 65nm CMOS. In *Solid-State Circuits Conference, 2007. ISSCC 2007. Digest of Technical Papers. IEEE International*, pages 98–589, feb. 2007.
31. X. Wu, J. Li, L. Zhang, E. Speight, R. Rajamony, and Y. Xie. Hybrid Cache Architecture with Disparate Memory Technologies. In *ISCA '09: Proceedings of the 36th annual international symposium on Computer architecture*, pages 34–45, New York, NY, USA, 2009. ACM.

AggMon: Scalable Hierarchical Cluster Monitoring

Erich Focht and Andreas Jeutter

Abstract Monitoring and supervising a huge number of compute nodes within a typical HPC cluster is an expensive task. Expensive in the sense of occupying bandwidth, and CPU power that would be better spend for application needs. In this paper, we describe a monitoring framework that is used to supervise thousands of compute nodes in a HPC cluster computer in an efficient way. Within this framework the compute nodes are organized in groups. Groups contain other groups and form a tree-like hierarchical graph. Communication paths are strictly along the edges of the graph. To decouple the components in the network a publish/subscribe messaging system based on AMQP has been chosen. Monitoring data is stored within a distributed time-series database that is located on dedicated nodes in the tree. For database queries and other administrative tasks a synchronous RPC channel, that is completely independent of the hierarchy has been implemented. A browser-based front-end to present the data to the user is currently in development.

1 Introduction

One of the often overlooked challenges in modern super-computing is the task to track system state, supervise compute node status and monitor job execution. As node counts increase, monitoring becomes a task that consumes a significant amount of network bandwidth and CPU power.

The development and investigation described in this paper were done as part of the TIMACS project and were sponsored by NEC and the German Bundesministerium für Bildung und Forschung.

E. Focht (✉) · A. Jeutter
NEC HPC Europe, Hessbrühlstr. 21b, 70565 Stuttgart, Germany
e-mail: efocht@hpce.nec.com; ajeutter@hpce.nec.com

M. Resch et al. (eds.), *Sustained Simulation Performance 2012*,
DOI 10.1007/978-3-642-32454-3_5,

Thus challenges for a HPC cluster monitoring system are:

- Minimise communication demands: bandwidth should be preserved for the application jobs.
- Scalability: keep growth-rate of infrastructure demands of the monitoring system well below the growth-rate of total compute nodes in the system.
- Minimise CPU usage: run as a subordinate task on the compute nodes but propagate critical system states as fast as possible.
- Fault tolerant and self recovery: a single failure of a compute node should not cause the monitoring system to collapse.

2 Previous Work

One of the first common and widely used tools to monitor large scale cluster hardware was the Berkeley University developed Ganglia [2]. Ganglia uses a distributed architecture approach and utilize unicast or multicast communication to send monitoring data to a master node. A configurable front end application displays the data in various ways and provides an overview of the whole system.

Van Renesse et al. developed and described in [10] an information management system that collects monitoring data and tracks system state on large computing sites. This system uses an hierarchical approach where compute nodes are put in zones. Zones are organized in a hierarchical fashion where each zone aggregates its data in relatively small portions to leverage bandwidth.

Marsh et al. investigated in [9] into scalability, reliability and fault tolerance of AMQP messaging systems. They proposed a federation hierarchy of nodes in conjunction with a dedicated configuration that is based on experimental data to gain maximum scalability.

Wang et al. described in [11] a messaging system using a publish/subscribe mechanism to send information over a distributed system. They added features to priorize topics and thus gained real-time performance for critical messages.

3 Architecture and Design

3.1 Core Design Decisions

The core design decisions for AggMon were driven by the target of reaching high scalability of the monitoring infrastructure while keeping the network as lightly loaded with monitoring data, as possible. Aiming at specialized HPC machines with huge numbers of compute nodes we consider a fixed or at least very slowly changing monitoring hierarchy to be a very realistic approach. O(1000–10000) specialized compute nodes deserve a hierarchy of dedicated administration nodes that take over the load of monitoring and keep as much as possible of it away from the compute nodes and compute network.

Scalability of the communication infrastructure for monitoring data is rarely addressed, but its choice can influence the way how to deal with increasing numbers of data reporters and temporary outages of the network or of administration servers. We decided for topic based publish/subscribe (eg. [8]) semantics. They allow for a nice asynchronous design of communicating components, the data producers can send out their data and forget about them, while data consumers can register handlers for the particular data topics they will be processing.

For collecting monitoring data we don't want to re-invent the wheel. Instead we want to be flexible and use data collected by already existing monitoring components like ganglia, nagios, collectd, collectl. In order to keep the traffic of monitoring data limited we use data aggregation heavily and only push aggregated data representing meaningful information about a group's state upwards the hierarchy tree.

The current value and time history of metric data is stored in distributed manner spread across the administration nodes, keeping the compute nodes free of the burden of disk I/O for monitoring data.

3.2 Hierarchy

In order to improve the scalability and manageability of huge computer systems a distribution of data and load is needed. Two approaches seem natural: the use of peer-to-peer and overlay networks or the use of a hierarchy or even combine both ideas like in [10].

We decided for a rather static hierarchy and against the use of overlay networks because for HPC systems the compute nodes should be kept free of any additional load which could spoil the scalability of the user applications. Therefore compute nodes should at most take care of generating their own monitoring data and sending it upstream the hierarchy path, but not need to "know" about monitoring data of other compute nodes or even try to aggregate data in any way. The hierarchy consists of groups and client nodes. Groups can contain groups or clients and must have at least one master. Group masters must not necessarily be located inside the group they are responsible for. The top level group is called "universe" and contains all groups and client nodes of the system. This hierarchy that can be represented as a direct acyclic graph can be compared to a UNIX filesystem hierarchy where groups correspond to directories and client nodes and masters correspond to files. The root path "/" corresponds to the "universe" group and a list of full file paths would describe all nodes and groups in the hierarchy.

Figure 1 shows a simple hierarchy consisting of eighteen compute nodes plus six master nodes spread over five groups. Each node in the system can be described by a unique path, for example:

$$/b/x/n07$$

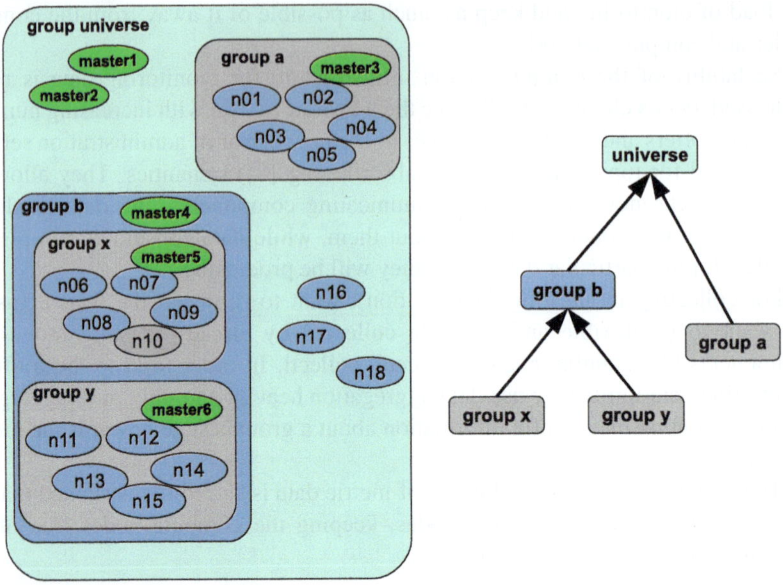

Fig. 1 Group hierarchy consisting of five groups where the groups *x*, *y* are contained in group *b*, and the groups *a* and *b* are contained in the group *universe*. The group dependency *graph* on the *right side* of the figure corresponds to the flow of group-aggregated monitoring data

is the full group path of node *n07*. It is a direct member of group *x*, which itself is member of group *b*. Each group's master node collects the monitoring information of its own nodes and of its subgroups. Time series information is kept on the group master nodes, aggregated monitoring data is pushed up the hierarchy tree to the higher level group's master nodes.

3.3 Components

The components built into the AggMon daemon are depicted in Fig. 2.

The core that links all components is the *channel*. It provides a topic based publish/subscribe abstraction. Channels can be opened, subscribed and published to. They are addressed with a URI.

The primary source of monitoring data are *importers*. They can run on compute nodes and collect local monitoring data or they can run on the group master nodes and collect metrics from the compute nodes. *Importers* publish the measured metrics to their group's master node with a topic that contains information on the data source, the data type and the originating host.

Each node that acts as a group master runs the *database* component. It subscribes to the group's channel and "sees" each metric published within the own group.

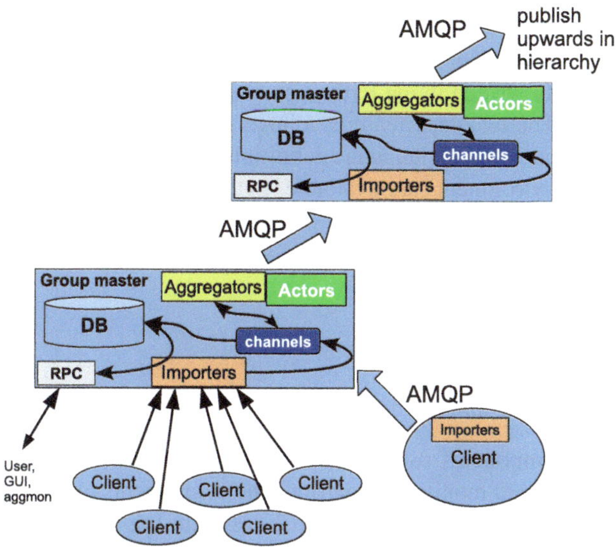

Fig. 2 Architecture or AggMon: Client nodes metrics are either published by importers running on clients or gathered by importers running on group master nodes. Group master nodes run various services like Database, Channels, Aggregators, and publish aggregated group data upwards in the monitoring hierarchy

Metrics are serialized and stored to disk. The time history of the monitoring data is also stored within the database. Each database instance keeps only information of its own group's nodes and subgroups. All group master nodes together form a specialized distributed monitoring database.

An *RPC* server component serves as user interface to the database component, allows querying the stored metrics and time-series information in synchronous way. Furthermore the RPC component is used to query and control the status of the AggMon daemon.

Aggregators are subscribers to the own group's channel and generate new metrics by using the information they see published by the group's members (nodes and subgroups). They can generate new host metrics (for example, compute number of fans running in a node out of the separately measured fan speeds) which are published on the group's channel as if they were measured by a normal importer. Or they can generate group metrics using any metrics from a group, on behalf of the group (for example, the highest node temperature in a group) and publish it upstream to the higher group's master.

Finally, *actors* are configured to execute commands initiated and triggered by aggregators or by the RPC component. This way an email could be sent out or an emergency script could be run if an aggregator considers that it discovered a problem that it should react to. Actors are planned but were not yet implemented at the time when this paper was written.

4 Implementation

AggMon is entirely written in the Python programming language [5] which allows fast prototyping. The language features a huge standard library with modules that can be loaded at runtime. The language definition is publicly available and currently several runtime environments are available. Python is very popular and supported by a large community of developers.

4.1 Publish/Subscribe

Message Brokers and Channels

All monitoring components use a topic-based publish/subscribe infrastructure to exchange the measured monitoring metrics. The components of AggMon act either as publishers, as subscribers or in some cases as both. Importers, the measurement components, collect data and publish it under particular topics on the network. Databases and aggregators act as subscribers to particular topics, aggregators publish the derived metric values.

The *channels* abstraction in AggMon is providing publish/subscribe semantics to the components. A channel is being addressed with a URL, its most generic form being:

```
<protocol>://[username:password@]hostname/channelname
```

Protocol is the used underlying implementation and can be one of:

- **local:** an optimization allowing threads local to a daemon process to exchange messages with topic based publish/subscribe.
- **amqp:** uses an external AMQP broker and is implemented with the py-amqplib library [4].
- **pika:** also uses an external AMQP broker, is implemented on top of the pika python library [3].

Username and password are used for authenticating with the message transport layer and are needed only when using one of the AMQP protocols.

The underlaying Advanced Messaging Queueing Protocol (AMQP, [1]) network relays the messages from the publisher to the subscriber and works as a message broker that can buffer data and decouples data generation and consumption. Its asynchronicity is very beneficial for scalability. AMQP has its origins in financial applications which deal with high numbers of transactions and strict requirements on availability and fault tolerance.

The AMQP broker is a component that should be considered a part of the underlaying network layer and is not subject to modification. The broker accepts messages that it receives from publishers. For each topic the broker maintains an internal queue where it appends the newly arrived message. Then the broker calls all subscribers for that particular topic and delivers the queued messages.

Usual brokers can run on single nodes but can also spread over networks and run on many nodes and build a big virtual clustered broker. Messages are automatically routed between broker instances to the ends where they need to be delivered to subscribers.

For our implementation we use the RabbitMQ messaging infrastructure [6], a widely used robust AMQP implementation written in the Erlang programming language.

Messages

The messages sent over the publish/subscribe infrastructure are JSON-serialized instances of the Metric class and consist of a set of key-value pairs. Some keys are mandatory:

- *name*: the metric name
- *source*: a hint on which component has created the metric
- *time*: the metric's measurement time
- *value*: the value of the measured entity.

Channel API

Applications can use the *channel* API to interface with the monitoring data network. The following code snippet sketches the usage of the three available API functions:

```
from aggmon.channel import *

# open a channel
channel = Channel.open( url, durable=False )

# subscribe to a topic, match group and metric name
topic = "group.*.*.metric_name"
channel.subscribe( topic, notify\_callback, raw... )

# publish a message on the channel
topic = "group.host.source.metric_name"
channel.publish( topic, message )
```

4.2 Importers as Metric Data Publishers

A publisher is a source of data it measures a physical value transforms and packs it into a Metric and publishes it under a certain topic in the network. The key point is that the publisher does not care and even does not know anything about the further treatment of the published data. One benefit of this scenario is that publishers can be separate short programs that are dedicated to a particular task. They even can sleep for longer periods and do not need to run permanently. One drawback of such asynchronicity is that common synchronous calls (RPCs, call and response pattern) are not possible. To overcome this the publisher can subscribe to a command topic and react on that. But most of the time this is not desirable. Another advantage of Public/Subscribe is that messages are guaranteed to be delivered. This is due to the possibility of the broker to store messages internally. Even if a subscriber is not running the messages are stored and delivered when the subscriber comes back online. Thus the grade of asynchronicity is only limited by the amount of data the broker can store. This provides fault tolerance and robustness.

4.3 Subscribers: The Metric Data Consumers

Subscribers are the contrary part of publishers. Subscribers connect to the broker and subscribe to a particular topic. The database and the aggregators are typical subscribers.

4.3.1 Database

The monitoring data is stored in a special purpose distributed database. Only group master nodes store data, they run an instance of the database for each group they represent.

The monitoring metrics are serialized to disk in a simple directory hierarchy. Each database instance for a particular hierarchy group path stores its data in an own directory. Each group member, host or subgroup, gets a subdirectory where each of its metrics is stored in a subdirectory of its own. Metrics attributes or metadata are stored in files named after each attribute. Time-series of metrics are abstracted into two classes: numerical records and log records and stored in separate files inside the metric subdirectory. Numerical records consist of the metric measurement time and value. Log records have string or unicode values and an additional optional "output" field. Currently each time-series file spans a certain time range, by default one day of data. In near future this will be extended by gradual thinning and averaging out of old data, in order to limit the amount of storage needed in a way similar to round-robin databases.

The database exposes several methods via RPC that can be called to retrieve data. All database instances are aware of the hierarchy and forward requests for data that is not available locally to remote database instances.

Database API

To query the database via RPC a connection must be opened to an arbitrary database instance. The query that would locally correspond to the call of a function

```
method(arg1, arg2, key3=arg3, ...)
```

is sent flattened as plain text in the form

```
method arg1 arg2 key3=arg3 ...
```

over the RPC channel. The database returns the results also as plain text that represents a valid Python object. This text can be evaluated via the *eval()* method to gain a Python object.

The concrete database API is still in development and being adapted and modified to the needs of programs that need to interact with the database, like a graphical user interface. Currently it consists of following functions:

- *dbSetOffline(group_path)*: Set current instance of database offline. In offline state the database doesn't commit received metrics to permanent storage but keeps them in memory in a log. Helper function for synchronizing database instances.
- *dbSetOnline(group_path)*: Commits all non-stored metrics from the log and sets the current instance of the database online.
- *getHostNames(group_path)*: List all hosts for which metrics are stored.
- *getLastMetricsByHostName(group_path, host_name)*: Return a MetricSet object that is a list of many Metric objects that are attributed to host_name. Note that the Metric objects do just contain one time-value pair, the most recent one! Other time, value records could be retrieved with getRecordsByMetricName.
- *getLastMetricByMetricName(group_path, host_name, metric_name)*: Retrieve the metric specified by host_name and metric_name. The metric contains the last time and value recorded.
- *getLastSeen(group_path, host_name)*: Return the last_seen timestamp and age for a host.
- *getMetricNames(group_path, host_name)*: List all metric names that are stored for a particular host name.
- *getRecordsByMetricName(group_path, host_name, metric_name, start_s, end_s, nsteps, step_s)*: Return a list that contains record objects. Each record has two attributes time_ns (time in 10E-9 s) and value. The argument start_s in seconds specifies the earliest record to be returned. No records newer than end_s (in seconds) are returned. Finally nsteps defines the number of steps (data points) to return. Like step_s this will lead to averaging for numeric data. The argument step_s gives the minimum time between two consecutive records. This method returns a list containing records.
- *getSummary(group_path, path)*: Returns A "directory" listing of a path inside the fs serialized metric database.
- *getTimeSeriesType(group_path, host_name, metric_name)*: Return the type of time-series stored for a metric on a host, i.e. it's class name. The returned string

contains the class name of the time series for the metric and is either "RRD" or "LOG".

- *findWhereMetric(group_path, metric_name, metric_attr, condition, value)*: Return hosts for which the given metric's attribute fulfills a particular condition. This method is implemented as a fast lookup that only scans the in-memory data and avoids expensive disk operations.
- *hierarchyGroupNames(group_path)*: Helper function that lists hierarchy group paths that are children to the passed group_path parameter. It helps recursing down the hierarchy tree without the need of having explicit knowledge of it.
- *getDBInstances()*: Lists database instances present on this node. Returns a list of group paths for which the current node is a master.

4.3.2 Aggregators

Aggregators are the components of AggMon that probably contribute mostly to its scalability. They are running on group masters and are subscribing to the group's metric channel. Two generic aggregator classes provide the skeleton for the concrete implementations: *Aggregator* and *DPAggregator*. Aggregator is a simple consumer that subscribes to only one topic and gets a channel passed in where to publish its derived metrics. DPAggregator is a dual-ported consumer, it subscribes to the topic of the metric it should aggregate and in addition it subscribes to its own metrics that might get pushed upwards from subgroups.

A set of aggregators were implemented on top of the two generic classes.

Host Aggregators

Host aggregators are actually creating a new, derived metric out of a measured one. The derived metric belongs to the same host as the old metric and is being published with the host's topic into the own group's channel. In that sense they don't actually aggregate data, but transform it. Two host aggregators are currently implemented:

- HostMaxAggregator: an example with little practical use. For a given metric it's largest value since the start of the aggregator is tracked and published. Could be useful, for example, for seeing the maximum swap space used on a node.
- HostSimpleStateAggregator: a complex aggregator that constructs nagios-like state metrics with the values: OK, WARNING, CRITICAL, UNKNOWN out of a measured metric of a host. It allows the definition of states and of conditions that must be fulfilled for the states. Useful, for example, for converting a numeric temperature metric into the more comprehensive states.

Group Aggregators

Group aggregators collect metrics from the own group and transform them into one derived metric on the behalf of the group, which is being published upstream on the hierarchy tree. These metrics very effectively reduce the amount of traffic and data exchanged inside the compute system's management network. At the same time they help finding quickly the problems in the system by descending the hierarchy tree: if the maximum of node temperatures is too large in the cluster this will be reflected by an aggregated metric in the universe group. Finding the exact source of trouble means: look one level deeper, find the subgroup or host belonging directly to the previous level which exceeds the critical temperature. If it is a host, the problem is found. If it is a subgroup, look though its members, recursively. This way only little information needs to be propagated to the root of the monitoring tree, and the detailed information is kept where it belongs to, on the group masters.

The following set of group aggregators have been implemented at the time of writing this paper:

- GroupMaxAggregator, GroupMinAggregator: publish a metric that contains the largest or smallest value of the original metric seen inside the group since the start of the aggregator.
- GroupMaxCycleAggregator, GroupMinCycleAggregator: publish a metric that corresponds to the largest or smallest value of the original metric seen in the latest cycle of measurements. A cycle is the time in which all group members (including subgroups) have published the original metric. In order to avoid waiting forever for lost group members the cycle time has an upper limit after which the aggregated metric is published in any case.
- GroupSumCycleAggregator, GroupAvgCycleAggregator: group cyclic aggregators that publish the sum or the average of the members' metrics seen within a cycle. Subgroup metrics are considered with their weight factor corresponding to the number of members they represent.
- GroupTristateCycleAggregator: aggregates nagios-like state metrics with values OK, WARNING, CRITICAL to a group metric having the value of the worst state seen within a cycle. It can give an immediate overview of the state of an entire group: OK or CRITICAL.

We are currently extending state aggregators to be able to trigger activities through configurable *actors* when states change.

4.4 Commands via RPC

A command channel is needed for different purposes within the monitoring framework. One reason is to send database queries to nodes running a database instance. Another reason is that components need adjustments, e.g., change the data collection interval. Since the publish/subscribe network is asynchronous it does not feature the required functionality. Within AggMon a common Remote Procedure Call (RPC) scenario is used to execute synchronous commands on remote nodes. Commands can also be emitted by a user interface (command line tool or GUI) and represent the data in a decent way.

4.4.1 Data Flow Within the Monitoring Framework

This example describes a system that gathers data from a Ganglia Monitoring System [2], publishes them with a dedicated topic on the Monitoring Framework before they got delivered to a database system to be stored for later retrieval by a command line tool.

The following components are involved in this scenario:

- Ganglia data collector: Ganglia has no mechanism to push data to the collector, thus the collector must actively retrieve data from Ganglia. Hence collection interval can be set by the collector. The collector establishes a connection (TCP/IP in this case) to Ganglia. Ganglia then sends XML formatted data on the socket and closes the connection. The Collector parses the XML data and generates several Metrics. Remember that a single Metric contains only a single measurement value. The final Metrics are published under a particular predefined topic to the AMQP broker.
- AMQP broker: The AMQP broker maintains a message queue each subscriber and topic. If a newly published message topic matches the queue topic of a subscriber the broker adds the message to the queue and invokes the subscribers notify function.
- Database: The database is a subscriber to a particular topic. It stores time-value pairs in a round-robin scheme. Augmented data like unit, source and origin of the data are also stored but overwrite previously delivered data.
- Command line tool: User interaction is a synchronous operation and thus uses the RPC channel to retrieve data from the database. Possible arguments are a particular type of metric or a time-frame within time-value pairs are to be retrieved.

This scenario shows, that the AMQP-based Messaging Framework as a central component decouples data generation at the source, data collection in the database and synchronous data retrieval via the command line tool. Another useful benefit is that developers can easily work on different components together due to the clearly defined interfaces. Usually the mentioned components are implemented as separate processes which increases reliability and adds some degree of fault tolerance (like previously discussed). It is further possible to stop and restart components and to add components during run-time.

5 Conclusion

The aggregating hierarchical monitoring component was built using publish/subscribe messaging and followed the corresponding distributed programming pattern. The inherent decoupling of the program's components as well as the use of Python have speeded up the development allowing for rapid prototyping and enforcing clean and clear interfaces between the components. The publish/subscribe approach is well extensible to other distributed system software for large computer systems, like, e.g., provisioning systems and parallel remote administration tools.

The rather static hierarchy approach for scaling the monitoring workload fits well HPC setups where the structure of the clusters is rather static as well, with dedicated administration nodes. The advantage of the static hierarchy is that compute nodes can be kept free of the heavier monitoring tasks like accumulating metrics, aggregating them and storing or retrieving them from a database. This reduces the potential for monitoring induced OS jitter while naturally enforcing a structure of the monitoring system with groups and group master nodes doing aggregation and holding the pieces of the distributed monitoring database. The contrary approach of P2P or overlay networks based monitoring would apply for rather dynamically managed cloud systems, with nodes being added and removed from the system very frequently. There performance and scalability of parallel programs is rather unimportant, therefore monitoring induced OS jitter can be tolerated.

The decision for using Python and AMQP had a positive impact on the development speed, but later it turned out that AMQP, due to its complexity, limits the message rate to an order of magnitude of 5–10,000 messages per second, while Python's fake multithreading limited the scalability of the daemon that was coded in parallel manner. This limits the performance of a daemon to about 3,500 metrics per second when using a small number of aggregators, but this is more than sufficient to serve the O(100) nodes which we envision to have in a group.

In order to further increase the performance we intend to add a channel implementation that works on top of ZeroMQ [7], a protocol similar to AMQP that has significantly lower overhead. In future we would also like to evaluate NoSQL databases as backend for storing the monitoring data, e.g., MongoDB. Those would allow for a much simpler connection with web frontends and further experiments with data aggregation executed directly on the database, e.g., with map-reduce algorithms.

References

1. AMQP: Advanced message queuing protocol. http://www.amqp.org/ (2012)
2. Ganglia monitoring system. http://ganglia.sourceforge.net/ (2012)
3. Pika. http://pika.github.com/ (2012)
4. py-amqplib. http://code.google.com/p/py-amqplib (2012)
5. Python programming language. http://www.python.org/ (2012)
6. RabbitMQ. http://www.rabbitmq.com/ (2012)
7. ZeroMQ: The intelligent transport layer. http://www.zeromq.org/ (2012)
8. Eugster, P.T., Felber, P.A., Guerraoui, R., Kermarrec, A.M.: The many faces of publish/subscribe. ACM Comput. Surv. **35**, 114–131 (2003). DOI http://doi.acm.org/10.1145/857076.857078. URL http://doi.acm.org/10.1145/857076.857078
9. Marsh, G., Sampat, A.P., Potluri, S., Panda, D.K.: Scaling advanced message queuing protocol (amqp) architecture withbroker federation and infiniband. In: OSU Technical Report. Department of Computer Science and Engineering, The Ohio State University, Columbus, OH 43210 (2009)

10. van Renesse, R., Birman, K.P., Vogels, W.: Astrolabe: A robust and scalable technology for distributed system monitoring, management, and data mining. Department of Computer Science, Cornell University, Ithaca, NY 14853 (2002)
11. Wang, Q., gang Xu, J., an Wang, H., zhong Dai, G.: Adaptive real-time publish-subscribe messaging for distributed monitoring systems. In: OSU Technical Report. IEEE International Workshop on Intelligent Data Acquisition and Advanced Computing Systems: Technology and Applications, Intelligence Engineering Lab., Institute of Software, Chinese Academy of Sciences P.O.Box 8718, Beijing 100080, China (2003)

Part III
Earthquake Modeling and Simulation on High Performance Computing Systems

Application of Vector-Type Super Computer to Understanding Giant Earthquakes and Aftershocks on Subduction Plate Boundaries

Keisuke Ariyoshi, Toru Matsuzawa, Yasuo Yabe, Naoyuki Kato, Ryota Hino, Akira Hasegawa, and Yoshiyuki Kaneda

Abstract In order to know why megathrust earthquakes have occurred in subduction zones such as the 2011 off the Pacific Coast of Tohoku Earthquake in Japan, we reconsider previous numerical simulation results and try to apply them to actual fields such as the 2004 Sumatra-Andaman earthquake and large interplate aftershocks of the 2011 Tohoku Earthquake. From this study, we propose that one of the possible reasons of pre-seismic change of the 2011 Tohoku Earthquake might have been smaller for its magnitude because its fault was composed smaller (*M 7* class) asperities including the off Miyagi earthquakes as occurred in 1978 and 2005. We also suggest that the next megathrust earthquake along Nankai Trough in southwest Japan may have detectable pre-seismic change because it is composed of three large (*M 8* class) asperities in Tokai, Tonankai and Nankai region. Our trial numerical simulation results by using vector-type super computer show that Dense Oceanfloor Network System for Earthquakes and Tsunamis (DONET) may be useful to detect the pre-seismic change of a possible *M 9* class coupled megathrust earthquake composed of Tokai, Tonankai, Nankai and Hyuga-nada asperities.

K. Ariyoshi (✉) · Y. Kaneda
Earthquake and Tsunami Research Project for Disaster Prevention, Japan Agency for Marine-Earth Science and Technology, Yokohama 236-0001, Japan
e-mail: ariyoshi@jamstec.go.jp; kaneday@jamstec.go.jp

T. Matsuzawa · Y. Yabe · R. Hino · A. Hasegawa
Research Center for Prediction of Earthquakes and Volcanic Eruptions, Graduate School of Science, Tohoku University
e-mail: matuzawa@aob.gp.tohoku.ac.jp; yabe@aob.gp.tohoku.ac.jp; hino@aob.gp.tohoku.ac.jp; hasegawa@aob.gp.tohoku.ac.jp

N. Kato
Earthquake Research Institute, The University of Tokyo
e-mail: nkato@eri.u-tokyo.ac.jp

M. Resch et al. (eds.), *Sustained Simulation Performance 2012*,
DOI 10.1007/978-3-642-32454-3_6,

1 Introduction

1.1 Spatial Distribution of Mega-Thrust Earthquakes

In world history (Fig. 1), most of mega-thrust earthquakes have occurred near oceanic trenches accompanied with tsunami, especially around Japan. On the basis of asperity map [38] or spatial distribution of source area of historical earthquakes [37], faults of megathrust earthquakes are thought to be composed of multi-segment. For examples, the 2004 Sumatra Andaman Earthquake is thought to rupture from Sumatra segment through Nicobar segment to Andaman segment [19]. The 2011 off the Pacific Coast of Tohoku Earthquake is thought to be composed of three main ruptured zones off Miyagi, far off Miyagi, and off Ibaraki [13], which generates tremendous tsunami in Miyagi, Iwate and Fukushima prefectures. In this chapter, we refer to the phenomenon that nearby earthquakes occur after time-lag significantly shorter than recurrence interval as "coupled earthquakes".

1.2 Modelling of Coupled Earthquakes

Ariyoshi et al. [1, 2] categorized the characteristics of the coupled earthquakes into two models: (i) Model of slip proportional to fault size, and (ii) Characteristic slip model. In the following sections, we review the two proposed model.

Figure 2 shows an example of model of slip proportional to fault size. For that slip amount (D) proportional to fault size has two types: "L-model" proportional to fault length (L) [32], and "W-model" proportional fault width (W) [28]. In case of proportional relation between W and L, both L-model and W-model are the same relation as similarity model with stress drop ($\Delta\sigma$) constant [14]. For coupled earthquakes, however, it is possible that the aspect ratio (L/W) of fault does not always keep constant as discussed later.

Figure 3 shows an example of characteristic slip model. Coupled earthquake keeps slip amount same as single earthquake. In other words, slip amount is independent of fault size (W and L) even if several earthquakes on the same fault occur simultaneously, which was observed in the North Anatolian fault [18] and the 1992 Landers earthquake [33].

1.3 Application to Actual Earthquakes

For inland earthquakes with lateral faults and high dip angle, the upper limitation of fault width tends to significantly lower than fault length [28]. This is probably because seismic slip decays abruptly in asthenosphere. It is thought that matured active faults often have deeper part of fault reaching asthenosphere (Fig. 4). This means that fault width cannot develop for deeper part any more, and keeps constant

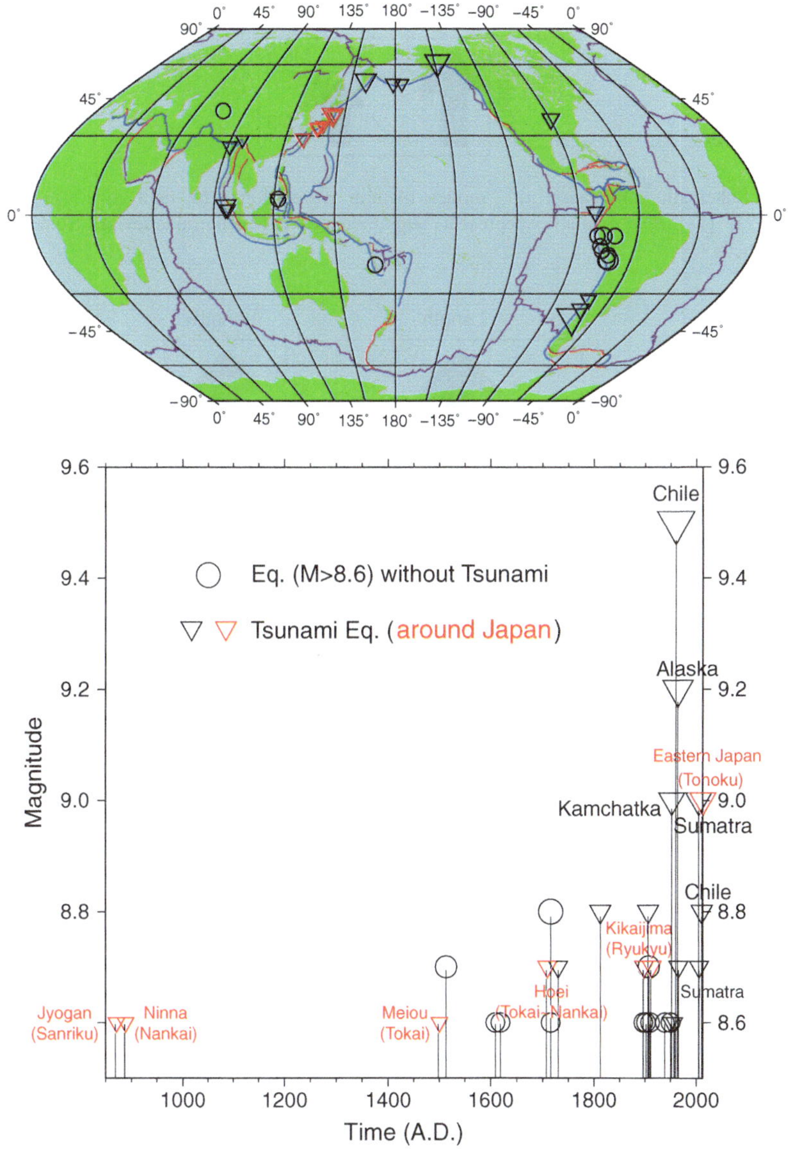

Fig. 1 (*top*) Map of epicenters and (*bottom*) time series for giant earthquakes ($M > 8.6$) originally revised from Global Significant Earthquake Database [22] with major plate boundaries (trench: *blue*, ridge: *purple*, transform: *orange*) determined by Coffin et al. [7]. "*triangledown*" represents tsunami earthquakes around (*red color*) and apart from (*black color*) Japan, while "*opencircle*" represents earthquakes without tsunami. The original magnitude is chosen from the available magnitude scales in this order: M_w, M_s, M_b, M_l, and M_{f_a}. In these figures, their database is revised on the magnitude for some earthquakes around Japan [2]

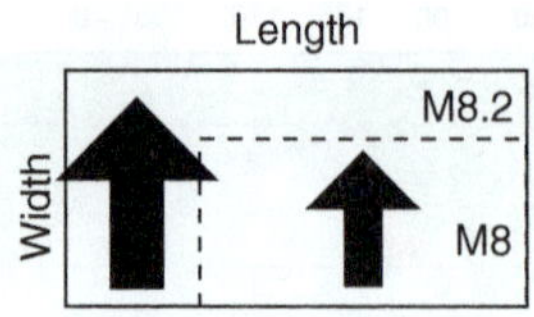

Fig. 2 Schematic illustration of "model of slip proportional to fault size" in case of $L = 2W$ [14]. *Bold arrow size* represents slip amount (D). This figure is revised from Ariyoshi and Kaneda [3]

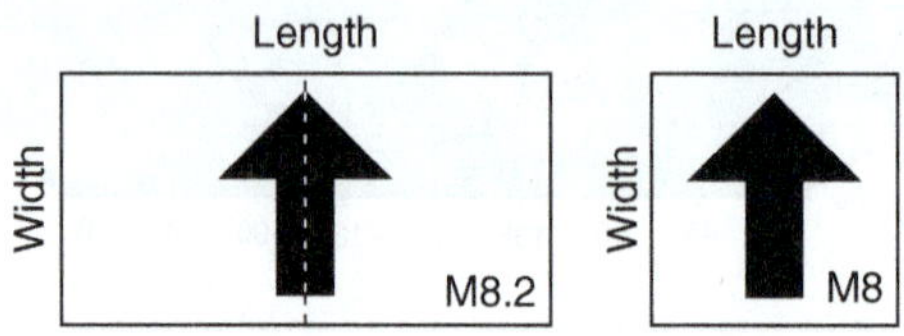

Fig. 3 Schematic illustration of characteristic slip model by comparing between coupled earthquake (*left*) and single earthquake (*right*). This figure is revised from Ariyoshi and Kaneda [3]

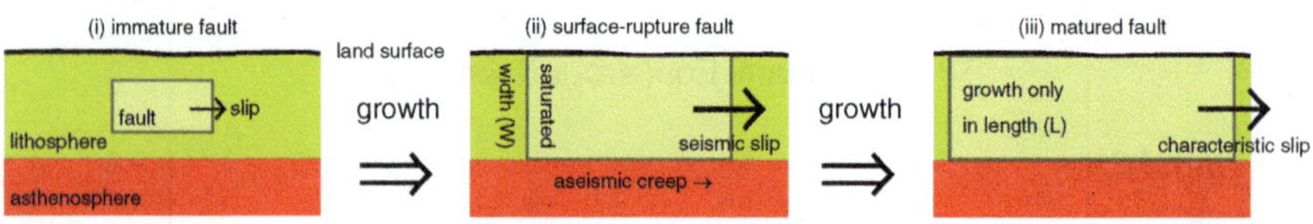

Fig. 4 Schematic illustration of fault growth process: (*i*) immature fault, (*ii*) fault with saturated width, and (*iii*) matured fault with the same slip amount as (*ii*). In case of "W-model", characteristic slip model is applicable between (*ii*) and (*iii*) because of failure of seismic slip to occur in asthenosphere. This figure is revised from Ariyoshi and Kaneda [3]

width in case of large earthquakes. Since keeping constant fault width makes constant slip amount for "W-model", "characteristic slip model" may be applied to matured active faults from the view of "W-model" [9].

For trench-type megathrust earthquakes, the upper limitation of fault width may be high enough to avoid saturation because of low dip angle. For examples, fault geometry (*width* $\times$ *length*) of the 2011 off the Pacific Coast of Tohoku Earthquake is thought to be 500 × 200 km [13], and sub-fault geometries of the 2004 Sumatra Andaman earthquake are 420 × 240 km for Sumatra segment, 325 × 170 km for Nicobar segment and 570 × 160 km for Andaman segment, respectively [19].

By treating the three sub-faults of the 2004 Sumatra Andaman earthquake as one giant fault, its aspect ratio does not keep constant as obeyed in the "model of slip proportional to fault size". On the other hand, the "characteristic slip model" would not quantitatively explain 20 m of its maximum slip amount.

In this chapter, we discuss the validity of both the "model of slip proportional to fault size" and "characteristic slip model" in some cases on the basis of numerical simulations, revealing the unknown characteristics and give a road map for earthquake prediction.

2 Numerical Simulation Studies

The two proposed model: "characteristic slip" and "slip proportional to fault size" is investigated from numerical simulation studies done by Ariyoshi et al. [1] and Kato [15], respectively. In this section, we review their calculation method and results, comparing slip behaviors and detectablity between them.

2.1 *Method of Earthquake-Cycle Simulations*

In order to focus specifically on the physical mechanisms of fault segment interaction, a planar plate interface is assumed in a homogeneous elastic half-space. The plate interface deeper than 103 km is assumed to slip at a constant rate of V_{pl} (relative velocity between the continental and oceanic plates) and the shallower part is divided into N cells. The slip for each cell is assumed to involve only a shear component in the dip direction and to obey a quasi-static equilibrium condition between the shear stress due to dislocation ($\tau_i^{dislocation}$) and frictional stress ($\tau_i^{friction}$). The stress is assumed to have both shear and normal components (σ_i) in the dip direction. The equations used in the simulation are as follows:

$$\tau_i^{dislocation}(t) = \sum_{j=1}^{N} K_{ij}(u_j(t) - V_{pl}t) - (G/2\beta)\frac{du_i}{dt}, \tag{1}$$

$$\sigma_i(t) = \sum_{j=1}^{N} L_{ij}(u_j(t) - V_{pl}t) + (\rho_r - \rho_w)gy, \tag{2}$$

$$\tau_i^{friction}(t) = \mu_i(t)\sigma_i(t), \tag{3}$$

$$\tau_i^{dislocation}(t) = \tau_i^{friction}(t), \tag{4}$$

Here the subscripts i and j denote the cell locations of an observation and a source, respectively. In Eqs. (1) and (2), K_{ij} and L_{ij} represent analytical Green's functions due to slip $u_j(t)$ in the jth cell for the shear and normal stress on the ith cell, respectively [23, 26]. The term $(u_j(t) - V_{pl}t)$ implies that we consider stress generated only by the amount of slip relative to the long-term average plate convergence [31]. The last term in Eq. (1) represents seismic radiation damping [27], where G and β are rigidity and shear wave speed, respectively. The last term in Eq. (2) represents the static effective normal stress assuming hydrostatic pressure, where ρ_r and ρ_w are the densities of rock and water, g is gravity acceleration, and y is depth. Equations (3) and (4) represent the frictional stress and the quasi-static equilibrium condition, respectively. The friction coefficient μ in Eq. (3) is assumed to obey a rate-and-state-dependent friction law [8, 29] given by

$$\mu_i = \mu_0 + a_i \log(\frac{V_i(t)}{V_0}) + b_i \log(V_0 \frac{\theta_i(t)}{d_{ci}}), \tag{5}$$

$$\frac{\mathrm{d}\theta_i(t)}{\mathrm{d}t} = 1 - V_i(t)\frac{\theta_i(t)}{d_{ci}}, \tag{6}$$

where a and b are friction coefficient parameters, d_c is the characteristic slip distance associated with b, θ is a state variable for the plate interface, V is slip velocity ($= \frac{\mathrm{d}u_i(t)}{\mathrm{d}t}$), and μ_0 is a reference friction coefficient defined at a constant reference slip velocity of V_0. The friction coefficient converges to a steady state value of $\mu_i^{ss} = \mu_0 + (a_i - b_i)\log(\frac{V_i}{V_0})$ when the slip velocity remains constant at V_i for a distance sufficiently longer than d_{ci} [29]. Therefore, μ_i^{ss} at velocity V_i is a function of $\gamma_i = (a_i - b_i)$, which represents frictional stability. If $\gamma_i > 0$, the slip is stable because frictional stress increases as the slip velocity increases, behaving like viscosity. If $\gamma_i < 0$, the slip is unstable and exhibits stick-slip behavior. The modeled spatial distributions of these frictional parameters are introduced in the next section. The six equations above are solved using the Runge–Kutta method with adaptive step-size control [25].

2.2 A Simulation of Characteristic Slip and Slip Proportional to Fault Size

In case of characteristic slip, Ariyoshi et al. [1] performed a simulation of Miyagi-oki earthquakes in a 2-D subduction plate boundary. Their simulation results show that pre- and post-seismic slip for following earthquakes of coupled earthquakes tend to be amplified significantly (about 2–4 times), while amplification of co-seismic slip is slightly (about 13–36 %). These results suggest that following earthquakes of coupled earthquakes with characteristic slip is more detectable than single earthquakes with same magnitude.

In case of slip proportional to fault size, Kato [15] formulated two adjacent large asperities reproducing the 1968 Tokachi-oki earthquake (M_w 8.2) breaking both the asperities and the 1994 Sanriku-oki earthquake (M_w 7.8) breaking one characteristic asperity [21]. Ariyoshi et al. [2] pointed out that seismic slip of the 1968 earthquake is approximately twice of the 1994 earthquake, considering that the ratio of seismic moment is $8.2/7.8 = 0.4$ and the asperity area ratio is almost 2 because both asperity has nearly same area. On the other hand, Kato [15] showed the simulation result that preseismic slip amount for the 1968 earthquake is nearly the same as the 1994 earthquake. These results suggest that a triggering nearby earthquake in case of "model of slip proportional to fault size" largely affects only on co-seismic slip, not on pre-seismic slip.

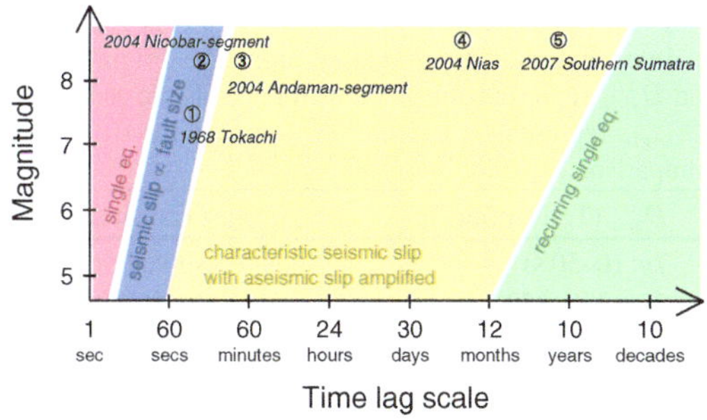

Fig. 5 A schematic relation between time lag, magnitude and slip model for the triggered earthquake of coupled earthquakes developed from Ariyoshi et al. [2]. Encircled numbers represent recently observed examples: (*1*) the 1968 Tokachi earthquake [21], (*2*) Nicobar segment of the 2004 Sumatra Andaman earthquake [19], (*3*) Andaman segment of the 2004 Sumatra Andaman earthquake [5, 19], (*4*) the 2004 Nias earthquake [6, 37], (*5*) the 2007 Southern Sumatra earthquake [17, 37]

2.3 *Relation Between Characteristic Slip with Slip Proportional to Fault Size*

Figure 5 shows a schematic relation between slip model and time lag of coupled earthquake on the basis of magnitude for the following (triggered) earthquake. As an example, we apply some coupled earthquakes regarding to the 2004 Sumatra Andaman earthquake into the relation in Fig. 5. Table 1 summarized slip components for the three sub-faults of the 2004 Sumatra earthquake which ruptured three seismogenic segments along trench—the Sumatra, Nicobar, and Andaman segments—and the average amount of co-seismic slip in each segment was roughly estimated at about $7m$, $5m$ and $<2m$, respectively [19]. The fault size factor of the coupled Nicobar and Sumatra segments relative to the Nicobar segment alone is about 2.35 (where the respective sizes (*trench* × *dip direction*) of the Sumatra and Nicobar segments are $(420 \times 240)\ \text{km}^2$ and $(325 \times 170)\ \text{km}^2$, respectively, and $(((420 + 325)\text{km} \times (240 + 170)\text{km}/(325\,\text{km} \times 170\,\text{km}))^{0.5} \sim 2.35)$ and the co-seismic slip amount of single-event earthquakes rupturing the Nicobar segment is estimated to be 2.7 ± 0.3 m based on the 1881 earthquake [5].

This implies that the co-seismic slip amount of the Nicobar segment is approximately proportional to fault size and, therefore, the stress drop model is preferable in describing interaction between the Sumatra and Nicobar segments. On the other hand, the observed co-seismic slip amount for the Andaman portion of the rupture is roughly the same as in the 1941 Andaman earthquake (2–$3m$ [5]) and post-seismic slip is substantial ($\sim 5m$, [19]), meaning that the characteristic slip model is preferable to account for interaction between the Nicobar and Andaman segments.

The 2005 Nias earthquake adjacent to the Sumatra segment occurred about three months after the 2004 Sumatra Andaman earthquake and was followed by the 2007 Southern Sumatra earthquake [17, 37] easterly adjacent to the source region of the

Table 1 Summary of slip components for the three segments ruptured by the 2004 Sumatra earthquake. Transit time represents time elapsed from the origin time of the 2004 Sumatra earthquake. D_{seis} and D_{slow} represent slip amounts for co-seismic and aseismic (slow) component, respectively [19]. D_{single} represents coseismic slip amount for single event based on previous researches [5] for comparison [2]

Segment	D_{seis} (Transit time)	D_{slow} (Transit time)	D_{single}
Sumatra	$7m$ (0–50 s)	Not resolved	Not known
Nicobar	$5m$ (230–350 s)	$5m$ (230–3,500+ s)	$2.7 \pm 0.3\,m$
Andaman	<$2m$ (350–600 s)	$5m$ (600–3,500+ s)	$2{-}3m$

Nias earthquake. Both of their post-seismic slips were amplified in western part [12], which were the same as directions of after-slip (post-seismic slip) arrivals [17, 37]. On the 2005 Nias earthquake (M_w 8.6), its seismic magnitude was largely equal to the 1861 Nias earthquake (M8.5) which occurred as single event, which is largely equal to the 2005. On the 2007 Southern Sumatra earthquake (M_w 8.5), its magnitude was neither as much as previous events occurred in 1833 (M8.9) nor moment magnitude expected from slip deficit [17]. These results also imply that the 2005 Nias earthquake and the 2007 Southern Sumatra earthquake were applied to the "characteristic slip model".

3 Discussion: A Question About the 2011 Tohoku Earthquake

Figure 6 shows asperity map of major trench-type megathrust earthquakes around Japan with spatial distribution of co-seismic slip for the 2011 off the Pacific Coast of Tohoku Earthquake. Focusing on the 2011 off the Pacific Coast of Tohoku Earthquake, we find that its source region covers some major asperities ($M7 \sim 8$) which overlap or neighbour each other. Unfortunately, it is well-known that pre-seismic change such as crustal deformation and seismicity has not been observed significantly. These observational results may suggest that the 2011 off the Pacific Coast of Tohoku Earthquake is composed of several asperities which behave as the "model of slip proportional to fault size" as mentioned above. In other words, pre-seismic change of the 2011 off the Pacific Coast of Tohoku Earthquake may be as small as $M7$ class earthquakes as observed for the 1968 Tokachi-oki earthquake [15]. In off Kanto and Boso, southward of the 2011 off the Pacific Coast of Tohoku Earthquake, there are several major source regions including the 1923 great Kanto earthquake (M7.9) and the 1953 off Boso peninsula earthquake (M7.4) which generated large tsunami. Since both the earthquakes are far away from the source region of the 2011 off the Pacific Coast of Tohoku Earthquake, both the southward major earthquake may behave as the "characteristic slip model". This suggests that their pre-seismic changes may be amplified so as to be detected by oceanfloor observations such as acoustic GPS [16] and/or repeating earthquakes[36].

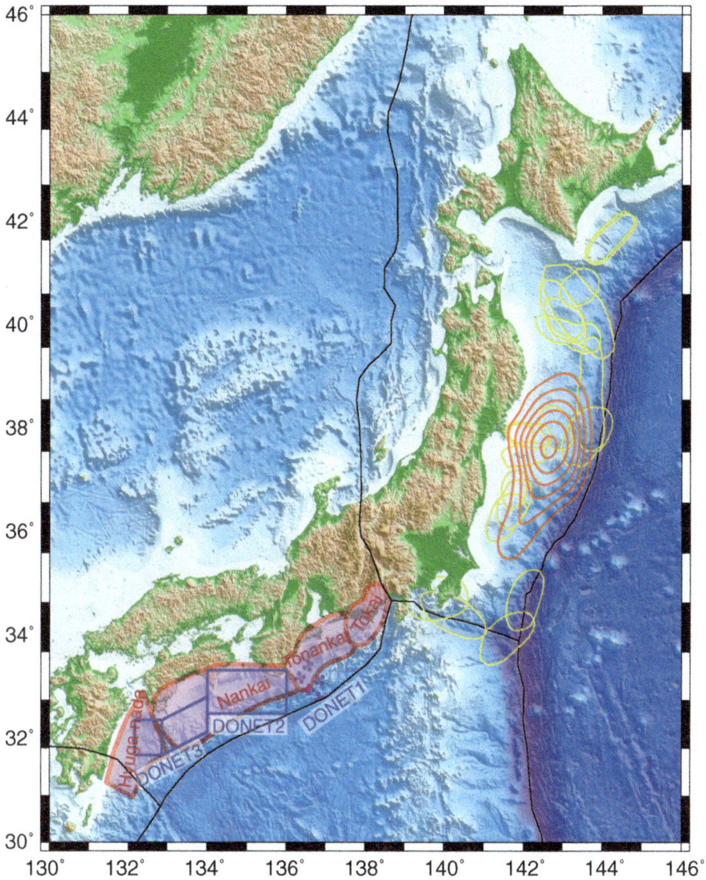

Fig. 6 Map of major asperities around Japan. *Black curves* are plate boundaries [24]. *Orange contours* represent co-seismic slip of the 2011 off the Pacific Coast of Tohoku Earthquake [11] as interval of 4 m. *Yellow ellipses* are the estimated source regions of past megathrust earthquakes (excluding outer-rise earthquakes) around the 2011 off the Pacific Coast of Tohoku Earthquake [34]. *Pink regions* from east to west along Nankai trough represent the seismogenic zones of Tokai, Tonankai, Nankai and Hyuganada earthquakes [35]. *Purple filled circles* and *open rectangle* regions represent observation points of DONET 1 and regions of DONET 2 and 3, respectively. This figure is modified from Ariyoshi and Kaneda [3]

Therefore, what we have to do is to develop:

- 3-D subduction plate model from geological surveys.
- Friction law based on rock laboratory experiments.
- Large-scale numerical simulations to combine above results.

These studies would determine the type of coupled earthquakes (slip proportional to fault size or characteristic slip) for major megathrust earthquakes including the 2011 Tohoku Earthquake.

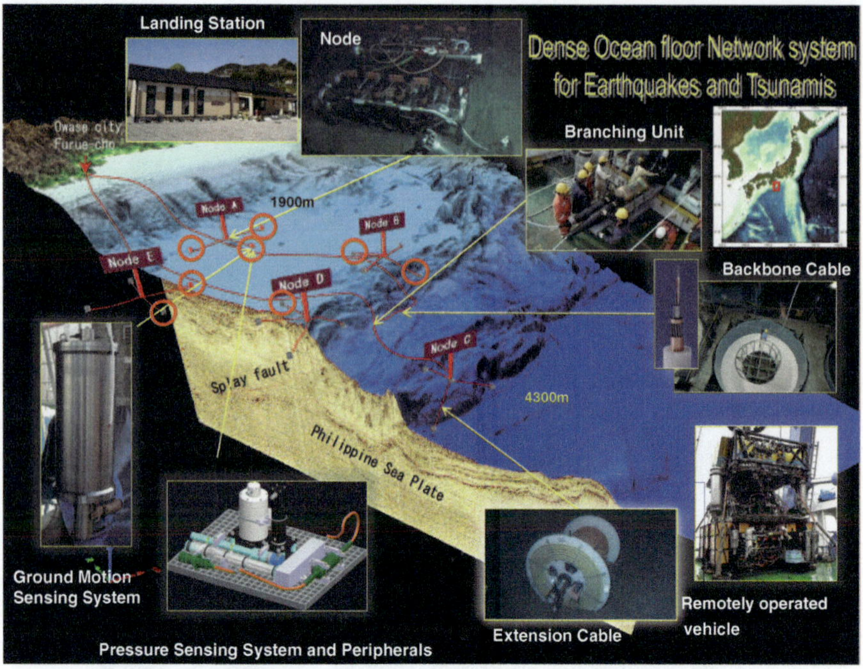

Fig. 7 An overview of Dense Oceanfloor Network System for Earthquakes and Tsunamis (DONET) in Tonankai region (DONET 1 in Fig. 8). This figure is modified from Ariyoshi and Kaneda [4]

4 Future Megathrust Earthquakes Around Japan

On the megathrust earthquakes along the Nankai Trough, it is thought that Tokai, Tonankai and Nankai earthquakes may occur in the near future and some researchers have pointed out that Hyuga-nada earthquake may be triggered by the $M9$ class coupled earthquakes composed of the three megathrust earthquakes [10]. However, size of asperities composing the possible $M9$ class coupled earthquakes along the Nankai Trough is significantly larger than that of the 2011 off the Pacific Coast of Tohoku Earthquake which may be composed of $M7$ class as shown in Fig. 6. This suggests that pre-seismic change of the possible $M9$ class coupled megathrust earthquakes along the Nankai Trough may be larger and is expected to be as large as the 1944 Tokai earthquake with detectable pre-seismic change reported by some researchers [20, 30]. Therefore, real-time monitoring of crustal deformation and seismicity is essential for us to detect the pre-seismic change in advance.

Figure 7 shows an overview of Dense Oceanfloor Network System for Earthquakes and Tsunamis (DONET) toward an anticipated Tonankai Earthquake. All of the twenty sets of preliminary interface have been installed just on July 31, 2011 and are to be prepared in consideration of the improvement of observation capability in the future.

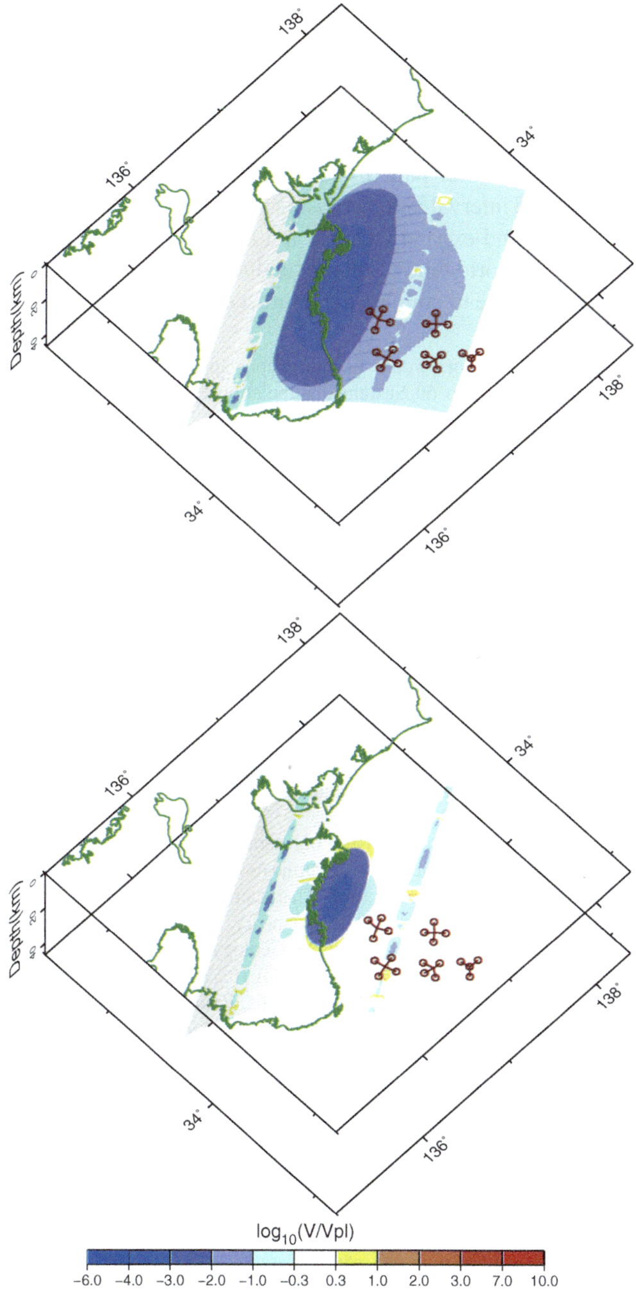

Fig. 8 Snapshots of slip velocity on the plate boundary about 20 years after the megathrust earthquake (*top*; interseismic period) and 2.5 years before (*bottom*; preseismic period). Twenty open circles with five nodes represent observation points of DONET 1 as shown in Figs. 6 and 7. This figure is modified from Ariyoshi and Kaneda [4]

Figure 8 shows examples of simulation results around Tonankai region, which suggests that monitoring the shallower part of slow earthquakes may be effective on the ground that it is more sensitive to the preseismic change of the megathrust earthquake because of free surface condition. In order to detect the preseismic slip of the next Tonankai earthquake in the near future, DONET would play an important role in monitoring shallower part of slow earthquake migration from the view of shortening recurrence interval and increasing migration speed.

We must develop and expand DONET not only in nationwide (DONET 2, 3 to be installed) but also worldwide of major subduction zones in order to mitigate the catastrophic disasters due to coupled megathrust earthquakes.

Acknowledgements The authors would like to deeply appreciate Professor H. Kobayashi for his inviting me to the Workshop on Sustained Simulation Performance 2012. The authors also thank DONET members and Tohoku University researchers. This study was partly supported by supercomputing resources at Cyberscience Center in Tohoku University and at Earth Simulator in JAMSTEC and by Grant-in-Aid (KAKENHI) for Young Scientists 23710212 and for Scientific Research on innovative Areas 20190449.

References

1. Ariyoshi, K., T. Matsuzawa, Y. Yabe, N. Kato, R. Hino, A. Hasegawa, and Y. Kaneda, Character of slip and stress due to interaction between fault segments along the dip direction of a subduction zone, J. Geodyn., 48, 55–67, doi:10.1016/j.jog.2009.06.001, (2009).
2. Ariyoshi, K., T. Matsuzawa, Y. Yabe, N. Kato, R. Hino, and A. Hasegawa, Consideration on the 2011 off the Pacific Coast of Tohoku Earthquake and the 2004 Sumatra Earthquake, JAMSTEC Rep. Res. Dev., 13, 17–33 (2011).
3. Ariyoshi, K. and Y. Kaneda, Characteristics of Interaction between Interplate Earthquakes from the view of Multi-scale Simulations, Nova Publication, in press (2012).
4. Ariyoshi, K. and Y. Kaneda, Frictional Characteristics in Deeper Part of Seismogenic Transition Zones on a Subduction Plate Boundary, Earthquake Research and Analysis - Seismology, Seismotectonic and Earthquake Geology, Sebastiano D'Amico (Ed.), ISBN: 978-953-307-991-2, InTech, pp. 402, 105–124, doi:10.5772/28884, (2012).
5. Bilham, R., R. Engdahl, N. Feldl, and S.P. Satyabala, Partial and complete rupture of the Indo-Andaman plate boundary 1847–2004, Seismo. Res. Lett., 76(3), 299–311 (2005).
6. Briggs, R. W., K. Sieh, A. J. Meltzner, D. Natawidjaja, J. Galetzka, B. Suwargadi, Y. Hsu, M. Simons, N. Hananto, I. Suprihanto, D. Prayudi, J. Avouac, L. Prawirodirdjo, and Y. Bock, Deformation and slip along the Sunda megathrust in the great 2005, Nias–Simeulue earthquake, Science, 311, 1897–1901 (2006).
7. Coffin, M.F., L.M. Gahagan, and L.A. Lawver, Present-day Plate Boundary Digital Data Compilation, University of Texas Institute for Geophysics Technical Report, 174, 5 (1998).
8. Dieterich, J.H., Modeling of rock friction: 1. Experimental results and constitutive equations, J. Geophys. Res., 84, 2161–2168 (1979).
9. Fujii, Y. and M. Matsu'ura, Regional Difference in Scaling Laws for Large Earthquakes and its Tectonic Implication, Pure Appl. Geophys., 157, 2283–2302 (2000).
10. Furumura, T., K. Imai, and T. Maeda, A revised tsunami source model for the 1707 Hoei earthquake and simulation of tsunami inundation of Ryujin Lake, Kyushu, Japan, J. Geophys. Res., 116, B02308, doi:10.1029/2010JB007918, (2011).

11. Geospatial Information Authority of Japan, The 2011 off the Pacific coast of Tohoku Earthquake: Postseismic Slip Distribution Model (Preliminary), http://www.gsi.go.jp/cais/topic110315.2-index-e.html, (2011)
12. Hsu, Y., M. Simons, J. Avouac, J. Galetzka, K. Sieh, M. Chlieh, D. Natawidjaja, L. Prawirodirdjo, and Y. Bock, Frictional afterslip following the 2005 Nias–Simeulue earthquake, Sumatra, Science, 312, 1921–1926 (2006).
13. Japan Meteorological Agency, The 2011 off the Pacific coast of Tohoku Earthquake 15 th report (in Japanese), http://www.jma.go.jp/jma/press/1103/13b/kaisetsu201103131255.pdf, (2011).
14. Kanamori, H. and D.L. Anderson, Theoretical basis of some empirical relations in seismology, Bull. Seism. Soc. Am., 65, 1073–1095 (1975).
15. Kato, N., Numerical simulation of recurrence of asperity rupture in the Sanriku region, northeastern Japan, J. Geophys. Res., 113, B06302, doi:10.1029/2007JB005515, (2008).
16. Kido, M., H. Fujimoto, S. Miura, Y. Osada, K. Tsuka, and T. Tabei, Seafloor displacement at Kumano-nada caused by the 2004 off Kii Peninsula earthquakes, detected through repeated GPS/Acoustic surveys, Earth Planets Space, 58, 911–915 (2006).
17. Konca, A.O., J. Avouac, A. Sladen, A.J. Meltzner, K. Sieh, P. Fang, Z. Li, J. Galetzka, J. Genrich, M. Chlieh, D.H. Natawidjaja, Y. Bock, E.J. Fielding, C. Ji, and D.V. Helmberger, Partial rupture of a locked patch of the Sumatra megathrust during the 2007 earthquake sequence, Nature, 456, 631–635 (2008).
18. Kondo, H., Y. Awata, . Emre, A. Doan, S. zalp, F. Tokay, C. Yildirim, T. Yoshioka, and K. Okumura, Slip Distribution, Fault Geometry, and Fault Segmentation of the 1944 Bolu-Gerede Earthquake Rupture, North Anatolian Fault, Turkey, Bull. Seism. Soc. Am., 95, 1234–1249 (2005).
19. Lay, T., H. Kanamori, C.J. Ammon, M. Nettles, S.N. Ward, R.C. Aster, S.L. Beck, S.L. Bilek, M.R. Brudzinski, R. Butler, H.R. DeShon, G. Ekstrm, K. Satake, and S. Sipkin, The great Sumatra-Andaman earthquake of 26 December 2004, Science, 308, 1127–1133, doi:10.1126/science.1112250, (2005).
20. Mogi,K., Two grave issues concerning the expected Tokai Earthquake, Earth Planets Space, 56, li–lxvi (2004).
21. Nagai, R., M. Kikuchi, and Y. Yamanaka, Comparative study on the source processes of recent large earthquakes in Sanriku-oki region: The 1968 Tokachi-oki earthquake and the 1994 Sanriku-oki earthquake, J. Seismol. Soc. Jpn., 54, 267–280, in Japanese, (2001).
22. National Geophysical Data Center, Global Significant Earthquake Database, http://www.ngdc.noaa.gov/hazard/earthqk.shtml, (2011).
23. Okada, Y., Internal deformation due to shear and tensile faults in a halfspace, Bull. Seism. Soc. Am., 82, 1018–1040 (1992).
24. Peter, B., An updated digital model of plate boundaries: Geochemistry, Geophysics, Geosystems, 4(3), 1027, doi:10.1029/2001GC000252, (2003).
25. Press, W. H., B. P. Flannery, S. A. Teukolsky, and W. T. Vetterling, Numerical Recipes, 2nd. ed., Cambridge Univ. Press, New York (1992).
26. Rani, S., Singh, S.J., Static deformation of a uniform half-space to a long dip-slip fault. Geophys. J. Int. 109, 469–476 (1992).
27. Rice, J.R., Spatio-temporal complexity of slip on a fault, J. Geophys. Res., 98, 9885–9907 (1993).
28. Romanowicz, B., Strike-slip earthquakes on quasi-vertical transcurrent faults: inferences for general scaling relations, Geophys. Res. Lett., 19, 481–484 (1992).
29. Ruina, A., Slip instability and state variable friction laws, J. Geophys. Res., 88, 10,359–10,370 (1983).
30. Sato, H., Some precursors prior to recent great earthquakes along the Nankai trough, J. Phys. Earth, 25, S115–S121 (Suppl.) (1977).
31. Savage, J.C., A dislocation model of strain accumulation and release at a subduction zone, J. Geophys. Res., 88, 4984–4996 (1983).
32. Scholz, C.H., Scaling laws for large earthquakes; consequences for physical models, Bull. Seism. Soc. Am. 72, 1–14 (1982).

33. Sieh, K., The repetition of large-earthquake ruptures, Proc. Natl. Acad. Sci., 93, 3764–3771 (1996).
34. The Headquarters for Earthquake Research Promotion, Long-term Evaluation of seismic activity off Boso to Sanriku (in Japanese), http://www.jishin.go.jp/main/chousa/kaikou_pdf/sanriku_boso.pdf, (2002)
35. The Headquarters for Earthquake Research Promotion, Long-term Evaluation of occurrence potentials or subduction-zone earthquakes around Hyuga-nada and Nansei island (in Japanese), http://www.jishin.go.jp/main/chousa/04feb_hyuganada/index.html, (2004)
36. Uchida, N., A. Hasegawa, T. Matsuzawa, and T. Igarashi, Pre- and post-seismic slow slip on the plate boundary off Sanriku, NE Japan associated with three interplate earthquakes as estimated from small repeating earthquake data, Tectonophysics, 385, 1–15 (2004).
37. Wiseman, K. and R. Brgmann, Stress and Seismicity Changes on the Sunda Megathrust Preceding the 2007 Mw 8.4 Earthquake, Bull. Seism. Soc. Am., 101(1), 313–326 (2011).
38. Yamanaka, Y., Kikuchi, M., Asperity map along the subduction zone in northeastern Japan inferred from regional seismic data, J. Geophys. Res. 109, doi:10.1029/2003JB002683, (2004).

Earthquake and Tsunami Warning System for Natural Disaster Prevention

Akihiro Musa, Hiroaki Kuba, and Osamu Kamoshida

Abstract Japan is one of the most earthquake-prone countries in the world. Earthquakes and tsunamis have repeatedly claimed many human lives and properties. The Japan Meteorological Agency has been operating a computer system for preventing disasters from earthquakes and tsunamis. NEC Corporation was contracted to develop its second generation system in 1993.

In this paper we describe the overviews of the system, the configuration necessary for higher reliability, and the warning mechanisms concerning a damaging earthquake. In particular, the warning mechanisms have several advanced features, and the Earthquake Early Warning notifies the public about the approach of a strong earthquake before it actually arrives. In addition, the warning mechanisms can predict the possibility of tsunami generation within 3 min after a quake. Moreover, we show the system's activities around the time of the massive earthquake on March 11, 2011, and then describe the necessity for further functional improvements to the system.

1 Introduction

On March 11, 2011, a massive magnitude 9.0 earthquake, called the 2011 off the Pacific coast of Tohoku earthquake, occurred off the coast of the Tohoku region of Japan. The strong quake hit nationwide from Hokkaido to Kyushu. Also high

A. Musa (✉)
Education and Science Solutions Division, NEC Corporation, 7-1 Shiba 5-chome, Minato-ku, Tokyo, 108-8001, Japan
e-mail: a-musa@bq.jp.nec.com

H. Kuba · O. Kamoshida
1st Government and Public Division, NEC Corporation, 7-1 Shiba 5-chome, Minato-ku, Tokyo, 108-8001, Japan
e-mail: h-kuba@ab.jp.nec.com; o-kamoshida@ab.jp.nec.com

M. Resch et al. (eds.), *Sustained Simulation Performance 2012*,
DOI 10.1007/978-3-642-32454-3_7,

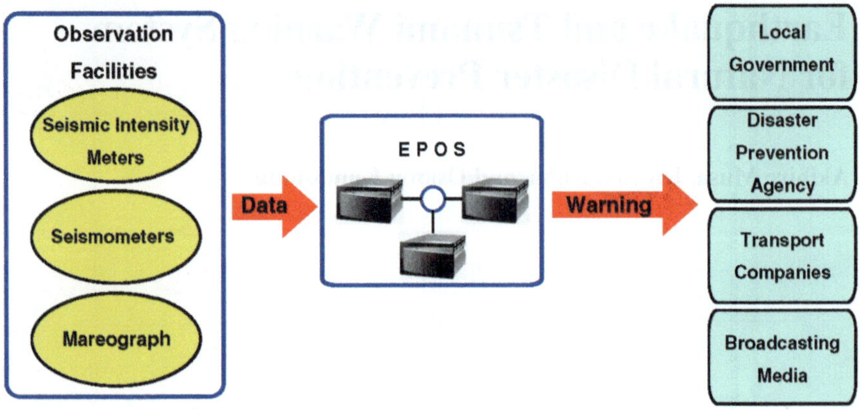

Fig. 1 Dissemination of earthquake early warning and tsunami warning

tsunamis caused by the earthquake struck the Pacific coast of Japan. The earthquake and tsunamis caused a lot of damage to the Tohoku and Kanto regions. Japan has often suffered from natural disasters such as earthquakes and tsunamis.

In 1875, the Japan Meteorological Agency (JMA) was established for the prevention and mitigation of natural disasters, and started its observations of earthquakes and weather. The JMA is the only government agency in Japan for forecasting natural phenomena related to the atmosphere, ocean, and earth, and it operates a number of observatories and weather stations across the country [1]. The JMA operates a seismic network for collecting seismic waveform data and a seismic intensity network for collecting seismic intensity data to monitor earthquakes. Each network comprises about 200 seismometers and about 600 seismic intensity meters throughout the country [2]. Moreover, the JMA uses seismometers and seismic intensity meters operated by local governments, the National Research Institute for Earth Science and Disaster Prevention (NIED), and universities.

The JMA has been using a computer system called the Earthquake Phenomena Observation System (EPOS) to help prevent earthquake and tsunami disasters. The EPOS observes earthquakes and tsunamis and issues Earthquake Early Warnings and Tsunami Warnings to the public as quickly as possible, as shown in Fig. 1. The JMA planned to conduct computerized seismic analysis for reducing warning processing times since 1983, and then developed the systems in 1987. NEC Corporation was contracted to develop its second generation system in 1993, and the EPOS is currently a fourth generation system. An overview of the EPOS is described with its activities around the time of the March 11, 2011 in this paper.

The rest of this paper is organized as follows. Section 2 presents the features of the EPOS, particularly the way Earthquake Early Warnings and Tsunami Warnings are issued. Section 3 describes the activities of the EPOS during the massive earthquake on March 11, 2011. Then, the plans to enhance the EPOS are shown for reducing the processing times and improving the prediction accuracy. Finally, Sect. 4 contains a summary.

2 Earthquake Phenomena Observation System (EPOS)

2.1 Hardware

A diagram of the EPOS is illustrated in Fig. 2. The EPOS contains seven subsystems and consists of NEC NX7700i and NEC Express5800 servers. The NX7700i is the enterprise server for mission critical systems. Therefore, it is used as the key server in each subsystem.

NX7700i/5020M consists of two cells, four I/O modules, and a crossbar network module. The cell contains four Intel Dual Core Itanium2 processors and a main memory, which are interconnected by NEC A3 chipset. The chipset is a cell controller with a 25.6 GB/s bandwidth, and it supports the detection function of the multi-bit error data and the retransfer function of the error data [3]. NX7700i/5012L and 5010E contain two Intel Dual Core Itanium2 processors, a main memory, and I/O sockets on their system boards.

The NX7700i servers have the following features for high reliability and availability:

- Dynamic Memory Resilience: to automatically deallocate faulty memory to prevent data corruption.
- Dynamic Processor Resilience: to enable taking a faulty processor offline without having to reboot the system.
- Redundancy and hot-swap functionality of disk, power, and fan units.

2.2 Duplicated Configuration

All the servers and network components in the EPOS are constructed in a fully duplicated configuration for higher reliability. As shown in Fig. 3 the primary and secondary servers work on the same processing in parallel. However, the interactive operations are executed on the primary server, and then, the results of the interactive operations are always stored onto the secondary server. The reason for this is that a complex management software is required to execute concurrently interactive operations in the two servers.

Every application process is activated as the child processes of the Process Controller in the server, and then the processes are always managed by it. If abnormalities with the child processes are detected, the Process Controller can restart the child processes using the following four functions:

- Files copy function: Related files are copied from the other server.
- Files initialization function: Related files are initialized.
- Files reused function: Only the process is restarted.
- Test function: Use trial data.

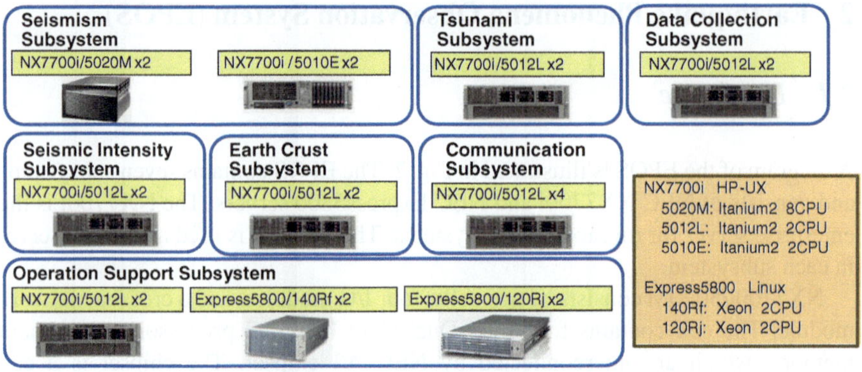

Fig. 2 Diagram of earthquake phenomena observation system

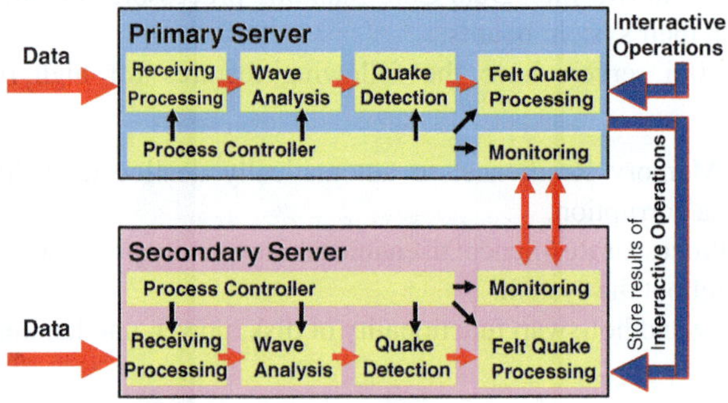

Fig. 3 Diagram of process in a duplicated configurations

Also, the secondary server becomes the primary server within two seconds in emergency situations.

Moreover, the JMA operates two EPOSs in Tokyo and Osaka for the prevention of disasters. The distance between Tokyo and Osaka is about 500 km, and therefore, if the Tokyo EPOS is damaged by an earthquake, the Osaka EPOS can continue its operations. The Osaka EPOS is a backup system of the Tokyo EPOS. Thus, the Osaka and Tokyo EPOSs have the same configuration and work on the same processing in parallel.

2.3 *Overview of Issuing Warning*

The EPOS has the six following processes for issuing Earthquake Early Warnings and Tsunami Warnings:

- Earthquake Early Warning Processing: to predict seismic intensities and to issue Earthquake Early Warnings.

- Seismogram Processing: to analyze seismic waves for determining the maximum amplitude and arrival times of waves.
- Hypocenter Determination Processing: to calculate the hypocenter and magnitude.
- Tsunami Processing: to predict tsunami occurrences.
- Tidal Processing: to observe tsunami waves for correcting Tsunami Warnings.
- Warning/Information Issue Processing: to issue Earthquake Information, Tsunami Warnings/Advisories, and Tsunami Information.

Earthquake Early Warnings and Tsunami Warnings are described as follows.

2.3.1 Earthquake Early Warning

Earthquake Early Warnings inform local governments, transport companies, people, and so on about a strong earthquake. The aim of the warning is to take countermeasures such as slowing down trains, stopping elevators, and enabling people to quickly protect themselves. For issuing the warning, the Earthquake Early Warning Processing uses the following characteristic features of an earthquake. The earthquake contains P waves (Primary waves) and S waves (Secondary waves). The propagation velocity of the P wave is faster than that of the S wave. Moreover, the velocity of a telegram is much faster that of the P wave. Meanwhile, the destructive power of the S wave is far greater than that of the P wave. Therefore, before the S wave arrives at an area, the processing issues a warning to that area.

Figure 4 shows a conceptual diagram of the processing. When a felt earthquake occurs, the processing automatically calculates the hypocenter and the magnitude of the quake using the P wave data detected near the hypocenter. Then, the seismic intensities are estimated in the areas around the hypocenter. If the estimated seismic intensities are above a 5-lower, a warning is issued to the given areas, where the estimated seismic intensity is 4 or greater, through the media, such as the TV, radio, and cell-phones. Here, the seismic intensity indicates the JMA seismic intensity scale, which has 10 degrees: 0 (imperceptible), 1, 2, 3, 4, 5-lower, 5-upper, 6-lower, 6-upper, and 7 [4]. The destructive power of a 6-lower or greater intensity may destroy many houses and buildings.

2.3.2 Tsunami Warning

When an earthquake occurs, the EPOS estimates the possibility of tsunami generation from the seismic waveform data. Japan is surrounded by the sea, and the coastline is long. Thus, the JMA subdivides the coastline into 66 regional blocks to predict tsunamis. If a damaging tsunami is expected, the EPOS issues Tsunami Warnings to the given coastal regions within 3 min after a quake. The Tsunami Warnings are categorized into three levels according to the estimated tsunami heights: *Tsunami Warning (Major Tsunami)*, *Tsunami Warning (Tsunami)*

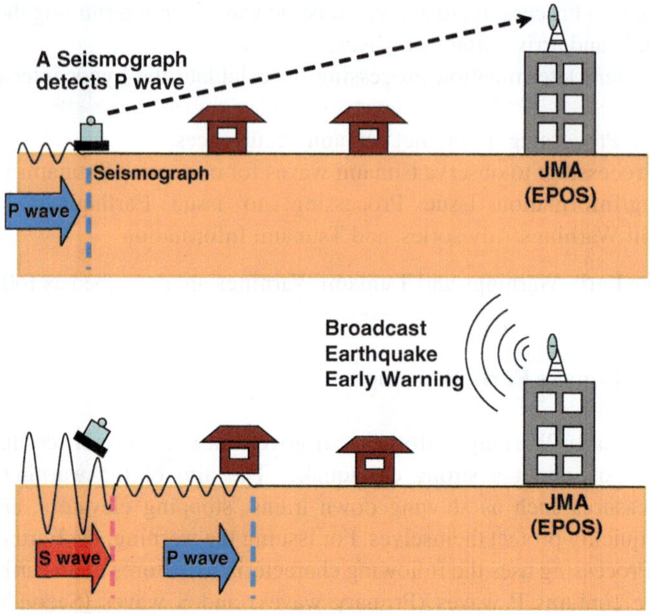

Fig. 4 Conceptual diagram of Earthquake Early Warning

and *Tsunami Advisory*. Each tsunami height of these warnings/Advisory is 3 m or more, up to 2 m, and about 0.5 m, respectively. Also, the EPOS issues Tsunami Information such as the estimated arrival times and tsunami heights in the given coastal regions. The warnings may be changed or updated based on the observed tsunami heights.

The height and arrival time of a tsunami are generally estimated by means of numerical simulations. However, since the simulations require a lot of processing time, Tsunami Warnings cannot be immediately issued. Thus, the JMA previously simulated tsunami propagations using about 100,000 earthquake scenarios involving different magnitudes, fault mechanisms, and various locations, as shown in Fig. 5. The simulation results were stored on a database, which is appropriately called the Tsunami Database. During tsunami processing, the EPOS selects the closest matching results from the Tsunami Database and determines the tsunami heights and arrival times for each coastal region [5].

There are various kinds of magnitudes worldwide. The magnitude used in the EPOS is usually calculated using two methods, the JMA magnitude [6, 7] and the moment magnitude [8]. The JMA magnitude has an advantage in that the processing time is within 3 min. Meanwhile, the processing time of the moment magnitude is within 15 min. However, the moment magnitude is a highly accurate calculation using a centroid moment tensor analysis. Therefore, the JMA generally makes a judgment about issuing Tsunami Warnings based on the JMA magnitude.

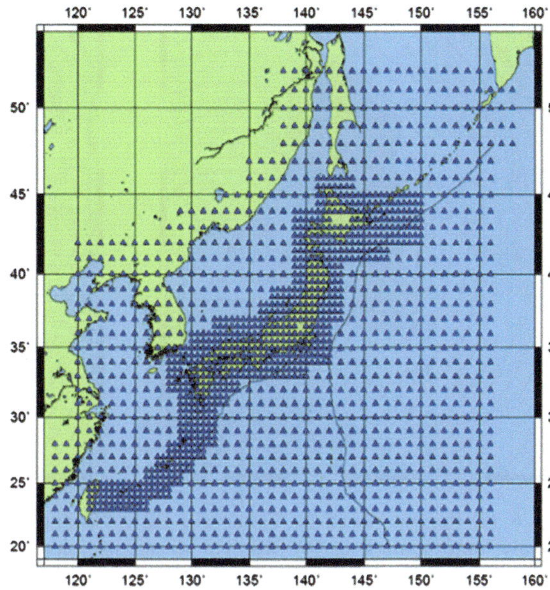

Fig. 5 Distribution of assumed faults

3 EPOS's Operations on March 11, 2011

The 2011 off the Pacific coast of Tohoku Earthquake occurred at 14:46:18 of March 11, 2011 (JST), the hypocenter of which was located 130 km ESE off the Oshika Peninsula and at a depth of 24 km [9]. The EPOS detected the seismic waves at 22 s after the quake and began several processes for issuing Earthquake Early Warnings and Tsunami Warnings. In this chapter, the operations of the EPOS on the quake are outlined, and then the necessity of further functional improvement is described.

3.1 Earthquake Early Warning

The Ishinomaki-Ouri seismograph station in Oshika Peninsula detected the P wave of the quake at 14:46:40, which was the first observation of the quake. The EPOS started the Earthquake Early Warning Processing, and the magnitude was calculated at 7.2. At 14:46:48, 22 s after the quake, the Earthquake Early Warning was issued to some areas of the Tohoku region, as shown in Fig. 6a. Specifically, the EPOS estimated that Miyagi prefecture experienced an intensity of a 4 or 5-lower, and Iwate and Fukushima prefectures had that of a 4. The circles in Fig. 6 show the fronts of the S-wave at the indicated times, which are the elapse times from the warning. Thus, the S-wave already arrived on the innermost circle as the warning was issued.

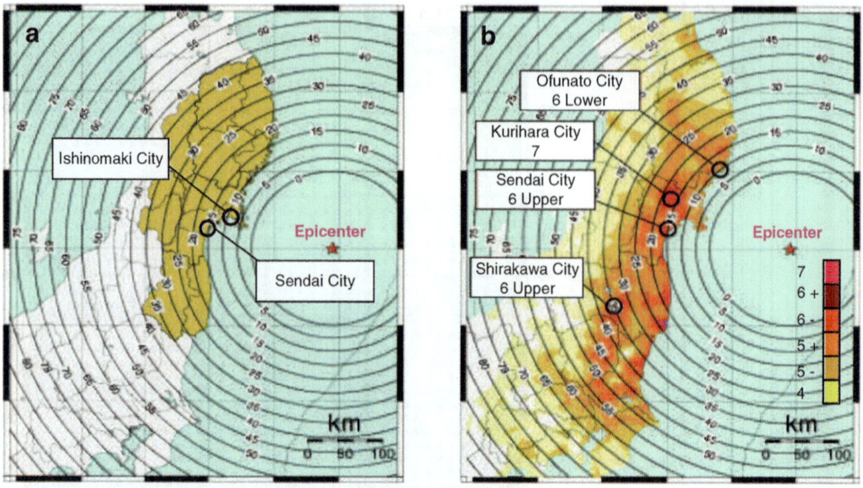

Fig. 6 (**a**) Regions issued the Earthquake Early Warnings (*yellow areas*) and arrival times of S-wave from the warning. (**b**) Observed seismic intensity and arrival times of S-wave from the warning

Figure 6b shows the observational results of the seismic intensity meters. The colored areas indicate the intensities of a 4 or greater. The observed intensities are greater than the predicted intensities. For instance, the EPOS predicted that the intensities at Sendai city and Kurihara city were a 5-lower. However, their observed intensities were 6-lower and 7, respectively. Moreover, the Tokyo area was not issued a warning, but the intensity in Tokyo was a 5-lower. This is because the predicted magnitude of a 7.2 was smaller than the actual magnitude. This shows that the estimated method of the magnitude in the Earthquake Early Warning Processing tends to be small for a large earthquake. Also, Fig. 6 indicates that Sendai city was hit by the S-wave at 15 s after the warning. The people near the epicenter did not have enough time to ensure the own safety from the quake.

3.2 *Tsunami Warning*

Since the earthquake occurred, the EPOS received a lot of seismic waveform data from all over the country, about 4,500 points. The EPOS calculated the hypocenter and the JMA magnitude, and then estimated the possibility of tsunami generation. Then, at 14:49, 3 min after the quake, the EPOS issued the Tsunami Warning shown in Fig. 7a. Here, the red line indicates the Tsunami Warning (Major Tsunami), the brown lines are the Tsunami Warnings (Tsunami), and the yellow lines show the Tsunami Advisories. Here, the predicted tsunami heights might be 6 m along the coast lines of Miyagi prefecture, and 3 m in Iwate and Fukushima prefectures. At this time, the EPOS calculated the magnitude at 7.9.

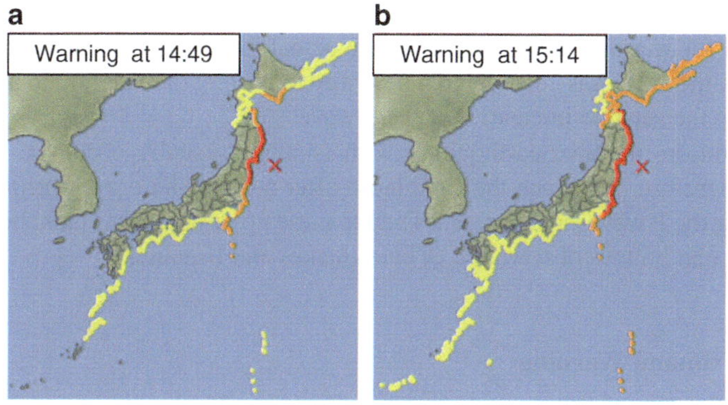

Fig. 7 Coast lines in Tsunami Warning

However, since the Tidal Processing observed abnormal tidal levels off the coast of Iwate prefecture, human operators who are tidal analysis experts re-determined the tsunami heights. Then, a Tsunami Warning (increased risk) was issued at 15:14, 28 min after the quake. The predicted tsunami heights might be 10 m or higher along the coast lines of Miyagi prefecture, and 6 m in Iwate and Fukushima prefectures. Moreover, the coast lines in the warning (Major Tsunami) were increased, as shown in Fig. 7b. Shortly after that, a massive tsunami of 8 m or higher struck the coast of Ofunato at 15:18. Furthermore, the coast between Iwate and Fukushima prefectures were struck by a massive tsunami.

The moment magnitude was not calculated within 15 min after the quake. This is because most of the seismometers in the country were off the scale and the EPOS waited to receive seismic waveform data from neighboring countries. Eventually, it was calculated that the moment magnitude was 8.4 at 16:00, 74 min after the quake. Therefore, the predicted tsunami heights in the first Tsunami Warning were lower than the actual tsunami heights.

3.3 *Enhancement Plan*

A reduction of processing times and the improvement of estimated accuracy in the EPOS have been in strong and continuous demand. Thus, the EPOS has been enhanced every year. After the 2011 off the Pacific coast of Tohoku Earthquake, enhancements in a massive earthquake have been particularly demanded.

3.3.1 Earthquake Early Warning

The method for the Earthquake Early Warning Processing can estimate a magnitude within only a few seconds. As shown in Sect. 3.2, however, the magnitude tends to be small for a large earthquake, and then the estimated seismic intensities and the areas for warning are smaller than the actual intensities and potential areas. In response, the JMA has considered a new way to correct the magnitude by means of observing the seismic intensities of the P waves.

In addition, massive earthquakes with tsunamis usually occur off the Pacific coast. Therefore, to reduce the time lapse after a quake before a warning requires observing the P waves in the ocean. The Japanese government has launched a project for the ocean bottom observation of earthquakes and tsunamis.

3.3.2 Tsunami Warning

Most seismometers in the country for the massive earthquake were off the scale, and the moment magnitude was not calculated. The JMA plans to install strong-motion seismometers, which can cover off the scale quakes of a magnitude 9.0, across the country. Moreover, the use of a supercomputer has been considered for reducing the processing times for the centroid moment tensor analysis.

In addition, the predicted tsunami heights now have a margin of error. Thus, highly accurate predictions require using observed data off the Pacific coast. The JMA has started considering using mareographs for automatically predicting tsunami heights.

4 Summary

Japan is one of the most earthquake-prone countries in the world. Earthquakes and tsunamis have repeatedly claimed many human lives and properties. The JMA has operated the EPOS since 1987 for preventing disasters from earthquakes and tsunamis. NEC Corporation has developed and enhanced the system since 1993.

The configuration for a higher reliability and the mechanism of warnings were described in this paper. Earthquake Early Warning Processing is one of the characteristic functions for this and it notifies people of the approach of a strong earthquake. A Tsunami Warning is normally issued within 3 min of a quake. However, the necessary improvements of the warnings for a massive earthquake were made glaring after the 2011 off the Pacific coast of Tohoku earthquake. Moreover, it has been predicted in Japan that massive earthquakes with magnitudes of 8.0 or more will occur off the Pacific coast within the next 30 years. Therefore, the JMA and NEC will attempt to develop a more sophisticated EPOS for the prevention and mitigation of massive earthquakes. In particular, NEC will take an important role in innovative technological solutions.

References

1. Japan Meteorological Agency., *The national meteorological service of Japan*, http://www.jma.go.jp/jma/en/Activities/brochure201003.pdf.
2. Japan Meteorological Agency., *Earthquakes and Tsunamis Disaster prevention*, http://www.jma.go.jp/jma/en/Activities/brochure_earthquake_and_tsunami.pdf.
3. Jun Yokoyama, Kenichi Suzuki, Kumiko Suzuki, Shinichi Kawaguchi., *NX7700i Series Supporting REAL IT PLATFORM*, NEC TECHNICAL JOURNAL, Vol.2, No.3, 13–17, 2007.
4. Japan Meteorological Agency., *Summary of Tables explaining the JMA Seismic Intensity Scale*, http://www.jma.go.jp/jma/en/Activities/intsummary.pdf.
5. Augustine S Furumoto, Hidee Tatehata, Chiho Morioka., *Japanese Tsunami Warning System*, The International Journal of the Tsunami Society, Vol.17, No.2, 85–106, 1999.
6. Akio Katsumata., *Revision of the JMA Displacement Magnitude*, Quarterly Journal of Seismology, Vol.67, No.1–4, 1–10, 2004.
7. Jun Funasaki., *Revision of the JMA Velocity Magnitude*, Quarterly Journal of Seismology, Vol.67, No.1–4, 11–20, 2004.
8. Thomas C. Hanks, Hiroo Kanamori., *A Moment Magnitude Scale*, Journal of Geophysical Research, Vol.84, No.B5, 2348–2350, 1979.
9. Japan Meteorological Agency., *The 2011 off the Pacific coast of Tohoku Earthquake -Portal-*, http://www.jma.go.jp/jma/en/2011_Earthquake.html.pdf.

Development of Radioactive Contamination Map of Fukushima Nuclear Accident

Akiyuki Seki, Hiroshi Takemiya, Fumiaki Takahashi, Kimiaki Saito, Kei Tanaka, Yutaka Takahashi, Kazuhiro Takemura, and Masaharu Tsuzawa

Abstract In order to continuously check the impact of radioactive substances on the health of residents and the environment, the Ministry of Education, Culture, Sports, Science and Technology (MEXT), universities and institutes have started conducting the research around Fukushima prefecture and neighboring prefectures. For the support of this research work, the Center for Computational Science and e-System (CCSE) at Japan Atomic Energy Agency (JAEA) has developed infrastructures which provide the information to the residents correctly and smoothly.

1 Introduction

Due to the massive earthquake and tsunami, the Fukushima Dai-ichi nuclear power plant has been damaged and has spread radioactive materials around the Fukushima site. More than 100 organizations started checking the radiation contamination status under the Strategic Funds for the Promotion of Science and Technology. The measurements were performed over the wide region not only in Fukushima prefecture but also in neighboring prefectures. Therefore, we could get massive data which are available as basic data. It is necessary to collect, analyze, and provide the information of radioactivity correctly and immediately.

For the purpose, we developed the Radiation dose Measurement Information Collection System (RMICS) for collecting measured data. The analysis software for the data from Kyoto University RAdiation MApping (KURAMA) system was developed to detect and collect error values in the enormous car-born survey data.

A. Seki (✉) · H. Takemiya · F. Takahashi · K. Saito
Center for Computational Science and e-Systems, Japan Atomic Energy Agency (JAEA), 5-1-5 Kashiwanoha, Kshiwa-shi, Chiba, 277-8587 Japan
e-mail: seki.akiyuki@jaea.go.jp

K. Tanaka · Y. Takahashi · K. Takemura · M. Tsuzawa
Japan Map Center, 4-9-6 Aobadai, Meguro-Ku, Tokyo 153-8522 Japan

M. Resch et al. (eds.), *Sustained Simulation Performance 2012*,
DOI 10.1007/978-3-642-32454-3_8,

Finally, those data can be seen by the public through the database system and the contamination map site which was also developed in this project.

In this paper, the background of this study is explained first. It is about reviewing the Tohoku earthquake and 2011 Strategic Funds. Then, the monitoring measurements are explained. The infrastructures which we developed for the research are then shown. Furthermore, some of the highlight data in this study are introduced. Finally, the summary of this paper and the future plans are explained.

2 Background

The Tohoku earthquake happened on March 11, 2011. It was 9.0 magnitudes at the Pacific Ocean from 100 km of northeastern coast and that scale of earthquake seldom happened in Japan. It generated tsunami which reached about 15 m at the Fukushima Dai-ichi nuclear power plant. It broke the power supply for cooling systems in the plant and the nuclear accident happened there.

Along with that, the radionuclides were released and spreaded throughout the wide region of Fukushima site. At the early phase following the accident, the airborne survey was conducted by the Department of Energy (DOE) in USA from March to April 2011. We got the result that the majority of radioactive plume went to the northwestern part from the nuclear plant. After that, the plume changed its direction to southwestern to the middle part of Fukushima prefecture.

In order to estimate the impact of the nuclear accident and to take appropriate countermeasures, MEXT commenced a project to produce radiation distribution maps under the 2011 Strategic funds for the Promotion of Science and Technology on June 6, 2011. For this project, JAEA and other institutes started various kind of radiation monitoring, from which radiation data were acquired for radiation maps. In addition, a database system including radiation data was developed by JAEA.

3 Radiation Monitoring and Mapping

The various kind of monitoring had been conducted in the project. Especially, the soil sampling, car-borne, air-borne survey results were displayed on the map. The soil sampling survey was required to determine the radionuclides in soil. The car-borne survey was effective in the continuous measurement of the air dose rate. As for the wide region survey, the air-borne survey results were also implemented.

3.1 Soil Sampling Survey

The left part of Fig. 1 shows the soil sampling area. In the beginning of this project, we divided the 2 km^2 and 10 km^2 regions to collect 5 samples in each region. 2 km^2

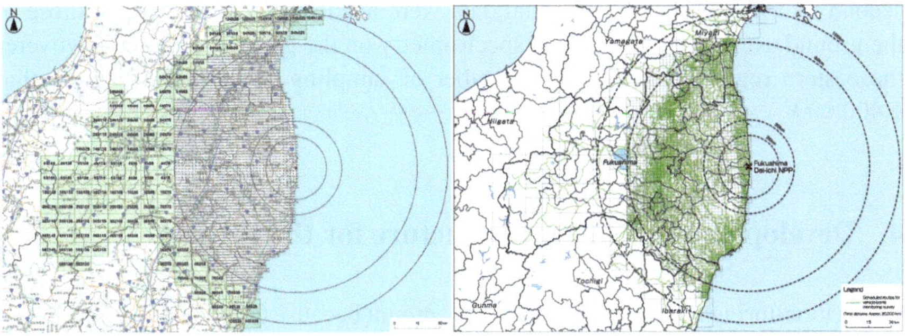

Fig. 1 The soil sampling area (*left*) and the car-borne survey area (*right*)

regions are in about 80 km radius from Fukushima nuclear site and 10 km^2 regions are over 80 km radius. The numbers of soil samples were greater than 10,000 which were collected by more than 400 people. We used special instrument for collecting the sample from the 5 cm surface layer. The soil was mixed in a plastic bag for even distribution and packed in the U8 plastic container for detection. After that, the concentrations of nuclide in those soil samples were measured by the Japan Chemical Analysis Center, the University of Tokyo, and other 19 organizations.

3.2 Car-Borne Survey

The right part of Fig. 1 shows the car-borne survey area. The green lines represent the roads which would be surveyed. We used the KURAMA system for the car-borne survey. That system measured the dose rates and GPS data at the same time. And those data were transferred to the storage server through the cellular network. KURAMA system was set in a car. The dose rate was converted to the rate outside because radiation dose was shielded by car. And for precise evaluation to the human body, the dose rate was also adjusted to the height of 1 m from the ground.

Eventually, the total mileage became more than 17,000 km but only covered the national and prefectural roads mainly within the Fukushima prefecture. Then, we have started next car-borne survey last December 2011 which commenced not only in Fukushima prefecture but also neighboring prefectures jointly with municipalities. The number of sampling points increased up to 140,000.

3.3 Air-Borne Survey

The air-borne survey was done under another research project conducted by MEXT. Those data have been also installed in database and map systems. Like car-borne survey, the amount of gamma-ray spectrum and GPS data were measured for every second by the helicopter. The aerial data were needed to be converted to those on the ground using data from in situ spectrometry on the ground. The survey covered the eastern region of Japan. The number of sampling points became more than 1,400,000.

4 Development of the Infrastructure for the Project

Those monitoring data became enormous amount because those data were measured over wide region in the eastern part of Japan. Furthermore various kinds of measurements were used for promoting the accuracy of environmental information. Those data have been required to open to the public promptly and accurately. Finally, those data should be used by different kinds of people.

To meet requirements, the infrastructures were developed for correcting, analyzing and providing those data. RMICS was developed for collecting data, KURAMA data analysis software for analyzing data, distribution map and database for providing data. Figure 2 shows images of relationship of those infrastructures.

4.1 RMICS

We developed RMICS for the collection of surveyed data at a sampling location. Figure 3 shows the image of RMICS. Before using the system, we recorded prior information manually, such as a place where soil sample has been collected, people who went to the location, what kind of survey meter they used. It was difficult to manage those data. Moreover, at the location where soil sample was collected, dose rate value and GPS data were written. The manual collection of data resulted to wrong information but with the help of RMICS, we could easily manage and restore precise data.

4.2 KURAMA Data Analysis Software

The KURAMA data analysis software has been developed too. The image of KURAMA data analysis software is shown in Fig. 4. KURAMA data become enormous data because one KURAMA system was recording a dose rate per 10 s. Following the amount of data, a lot of error values which are due to the detector and GPS were included in such KURAMA data. This software helps us to detect those error values and remove them quickly.

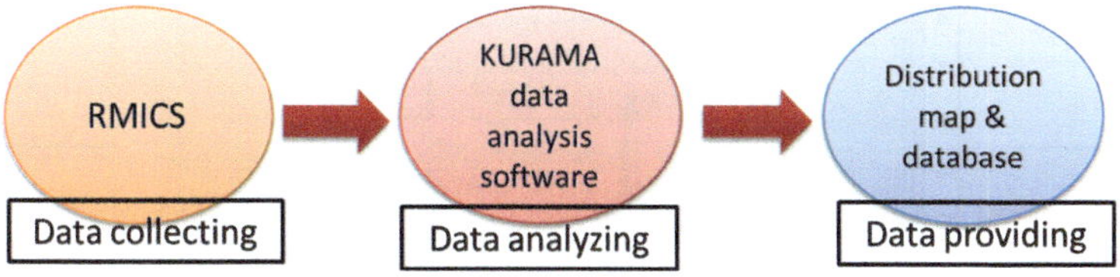

Fig. 2 The image of the relationship of the infrastructures which are developed in this project

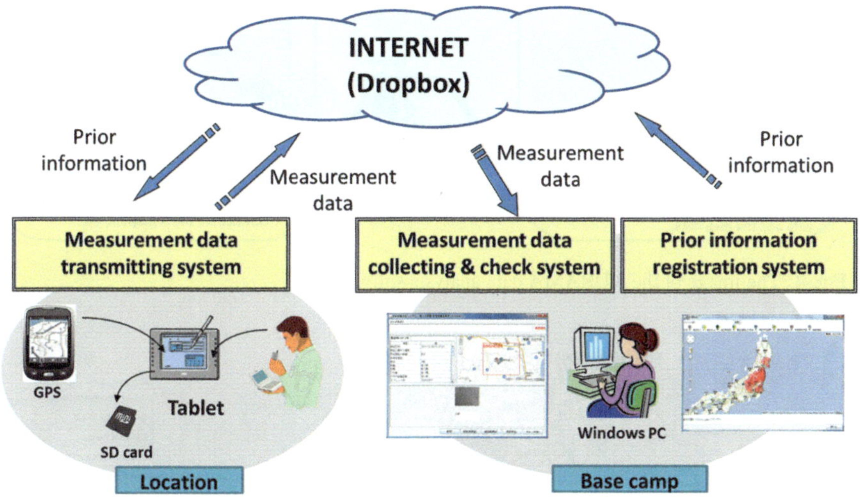

Fig. 3 The image of the Radiation dose Measurement Information Collection System

4.3 Distribution Map System

We prepared some infrastructures for providing radiation data to the public. One is the distribution map system. Those maps have been available through the internet since October in 2011. A user can recognize the dose rate in air (unit: Sievert per hour, Sv/h) and contamination of radioactivity in the soil (unit: Becquerel per square meter, Bq/m2) around the place of interest. There are DENSHIKOKUDO, PDF, and smartphone types of maps. The soil, car-borne, and air-borne survey data are available like Fig. 5. The functions of comparison of those data are also contained. Moreover DENSHIKOKUD map has the zooming-up function which shows dose rate at many locations in detail. And the map has information of places where elementary and junior high schools are located. The functions for searching the name of places or the latitude and the longitude are also available.

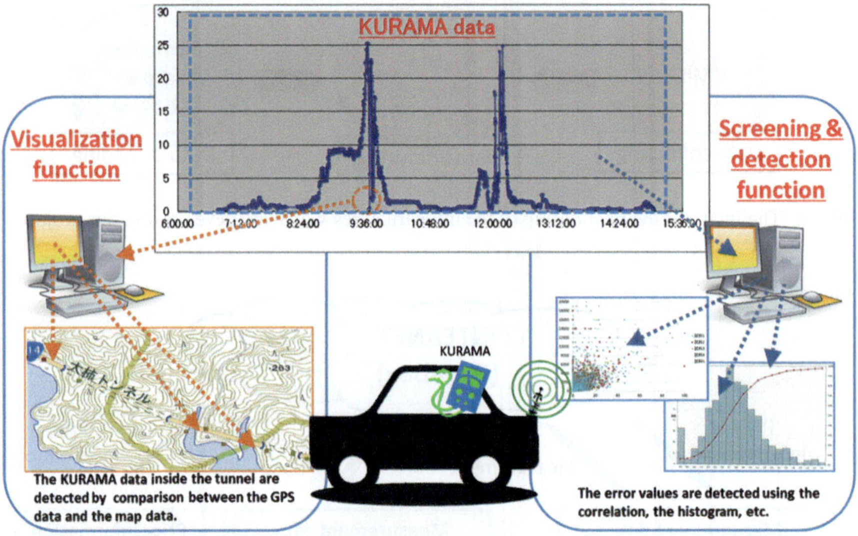

Fig. 4 The image of the KURAMA data analysis software

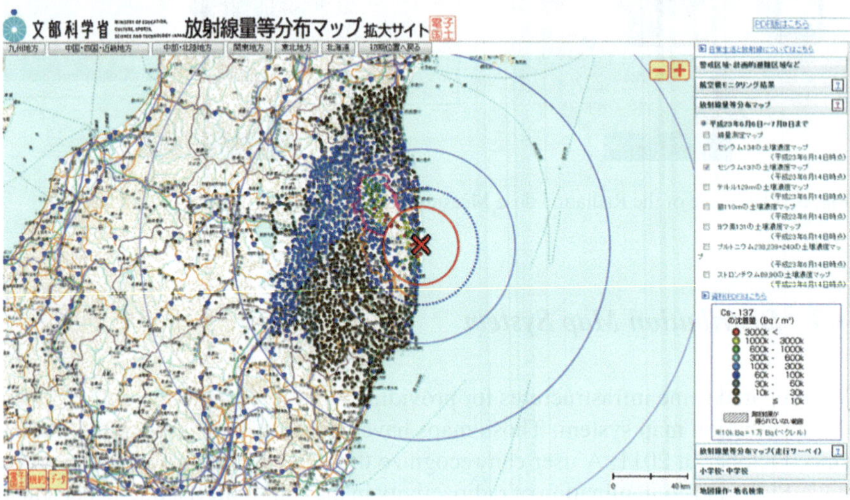

Fig. 5 One example of the distribution map system

4.4 Distribution Database System

The distribution database system which is available on the internet has been also developed to indicate data on the map in detail. A user can select the type of data, prefecture and municipality, then the data are displayed in the tabulate form, as shown in Fig. 6. Furthermore some detailed information such as soil's weight,

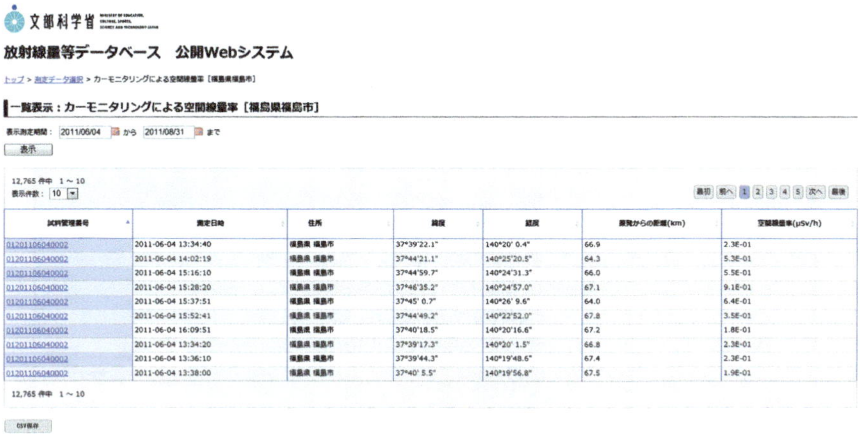

Fig. 6 One example of the distribution database system

survey meter can be confirmed in the table. The data can be download with the form of XML files and CSV files from the same web site. Then, the downloaded data can be useful for our works and researches.

5 Results

As the result of the first research project, the maps of Cs-134, Ce-137, I-131, Te-129m and Ag-110m had been made for about 2,200 points in Fukushima area. The obtained results are expected to be utilized as valuable data to examine the radioactive plume released initially from the nuclear power plant, ascertain how radioactive substances have been deposited on the ground surface, and assess people's exposure doses based on these results. The following introduces comparison of the distribution about Cs-137, I-131, Te-129m and Ag-110m in which the typical features are shown.

5.1 Distribution Map

Figure 7 shows a map of Cs-137 distribution in soil. The half-life of Cs-137 is 30 years. These colors show the different levels of concentration. The red color means more than 3000k Bq/m2. The amount of Cs-137 is larger than the other radioactive nuclides except for Cs-134.

The map of I-131 is also displayed in Fig. 8. The red color means more than 5000 Bq/m2. It is difficult to obtain the distribution of I-131 because its half-life is just 8 days. But, we were able to ascertain the distribution of I-131 as of June to July widely and in detail.

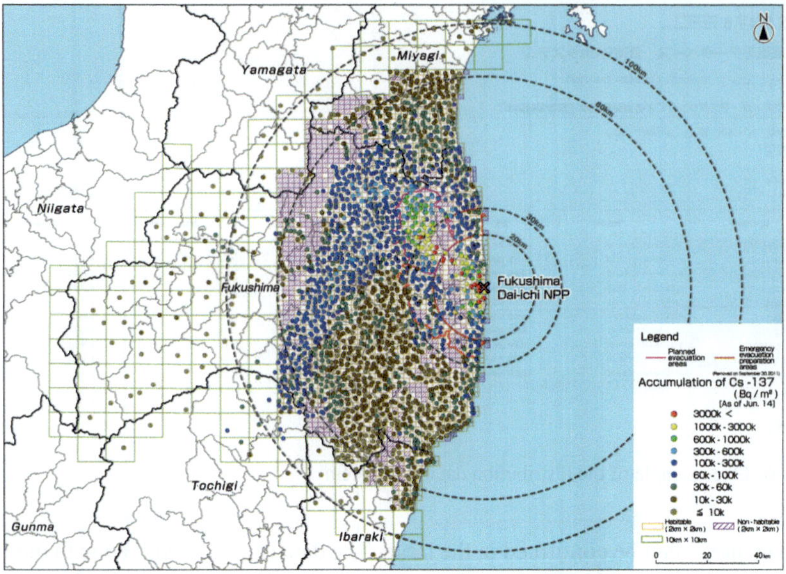

Fig. 7 The distribution map of Cs-137

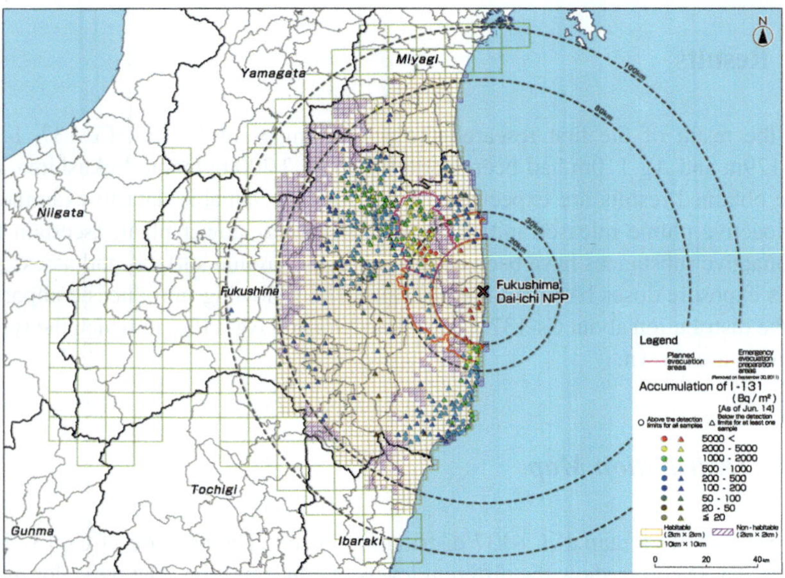

Fig. 8 The distribution map of I-131

5.2 *Distribution of the Ratio*

In order to check the deposition of radionuclides on the ground surface, we compare the ratio of their deposition amount against those of Cs-137. The ratio of deposition

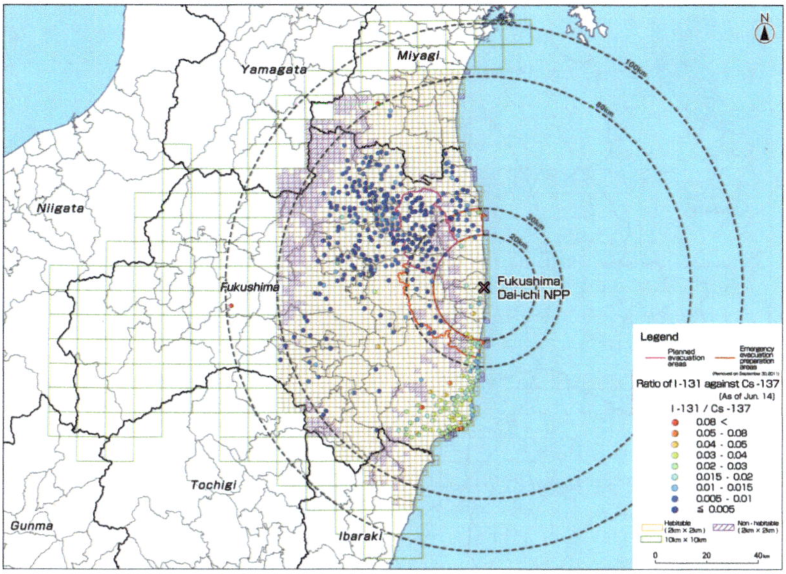

Fig. 9 The distribution of I-131/Cs-137 ratios

amount of I-131 against those of Cs-137 are shown in Fig. 9, Te-129m in Fig. 10, Ag-110m in Fig. 11. Relatively higher ratio of deposition amount of the nuclides to those of Cs-137 are confirmed along the coast areas compared with the surrounding areas. It is found that I-131 and Te-129m are deposited on the ground surface at the southern coastal areas.

Those results suggest the ratios of I-131, Te-129m, Ag-110m to Cs-137 contained in radioactive plumes and their chemical forms are different depending on each radioactive plume when released from the nuclear power plant. And weather conditions were not the same when these nuclides and Cs-137 were deposited on the ground surface.

5.3 Contributions of Dominant Radionuclides to the Total External Effective Dose

The contributions of dominant radionuclides to the total external effective dose were calculated from the radioactivity in the soil at the data of June 14 in 2011 for 45 selected locations. The radioactive cesium nuclides (Ce-134 and Ce-137) dominantly contribute to the effective dose, as summarized in Table 1. Thus, it can be understand that Ce-134 and Cs-137 should be taken into account for any environmental countermeasures such as decontaminations.

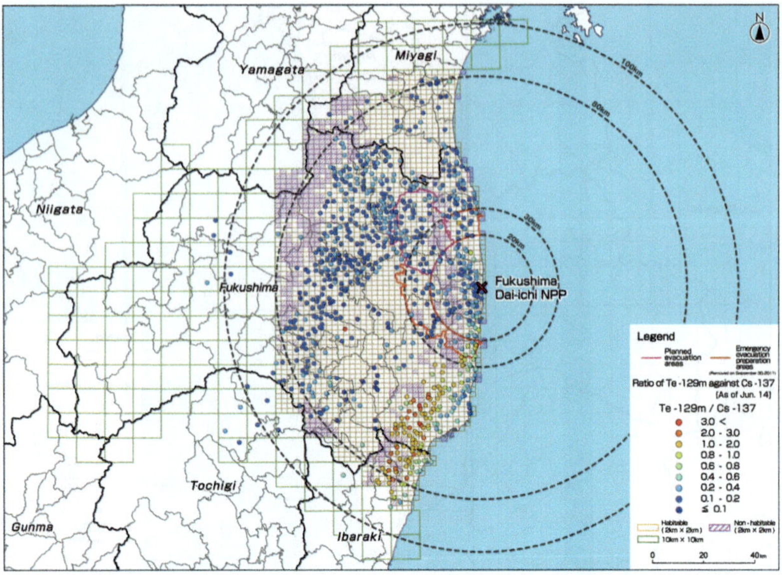

Fig. 10 The distribution of Te-129m/Cs-137 ratios

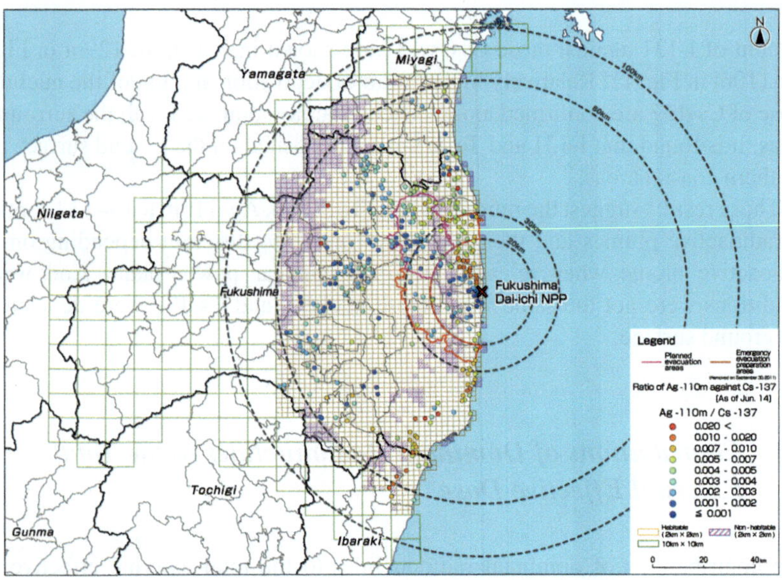

Fig. 11 The distribution of Ag-110m/Cs-137 ratios

Table 1 The contributions of dominant radionuclides to the total external effective dose

Nuclide	Ratio to the total effective dose rate
Cs-134	0.71
Cs-137	0.29
Te-129m	0.007
I-131	0.001
Ag-110m	0.001>

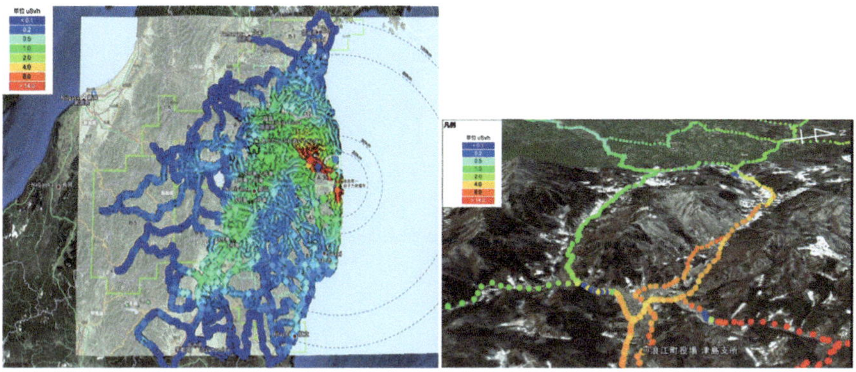

Fig. 12 The car-born survey result on the Google Earth

5.4 *Car-born Survey Data*

Figure 12 shows the car-born survey result on the Google Earth. In the left of Fig. 12, we can see that the plume went to northwestern part and after that, it changed its direction to southwestern part. The right of Fig. 12 shows the part of car-borne survey results in Namie town in Fukushima prefecture. The middle of this figure is the mountain. The radiation levels are different from each side. That means the radionuclides plume was passing through right side of the mountain.

6 Summary and Future Plans

The first mapping project has been finished by November 2011. JAEA developed the infrastructures for collecting, analyzing, and providing the surveyed data. With the use of those infrastructures, a lot of basic data for evaluations and countermeasures were corrected. Additionally, those data are used in providing the public through the database and map system.

In the future, the periodical surveys on contamination are continued. Actually, the second mapping project has started since December 2011. And we are going to construct the simulation models for the prediction of migration, the finding, the result of effective decontamination, and the evaluations of the effects on the human body. We believe those simulations will be accomplished by super computer.

Acknowledgements I want to thank the Visible Information Center, Inc. for constructing the database system and operating the database and map systems. I have furthermore to thank the Hitach East Japan Solutions, Ltd, for developing the RMICS and the KURAMA data analysis software.

Source Process and Broadband Waveform Modeling of 2011 Tohoku Earthquake Using the Earth Simulator

Seiji Tsuboi and Takeshi Nakamura

Abstract We have calculated broadband synthetic seismograms for March 11, 2011 Tohoku earthquakes using the Spectral-Element Method. We use finite source models by using a set of sub-events distributed along the fault surface, retrieved by inversion of body waves (Nakamura et al., 2010). The finite source model used in this simulation estimates Mw to be 9.1. The fault dimension is 460 km times 240 km with the source duration time of 150 s. We use the Earth Simulator2 of JAMSTEC to calculate preliminary synthetic seismograms for this finite source model. We used 726 processors of the Earth Simulator 2, which should provide synthetic seismograms that are accurate up to about 5 s and longer. The comparison of the synthetic seismograms with the observation for this event shows that synthetic P-waveforms model the observed seismogram quite well, reflecting that the finite source model is quite precise. This source model shows that the maximum slip occurs at depth of 20 km and propagates to shallower region, which is consistent with the fact that the tsunami excitation was significant for this event. Azimuthal dependence of misfits of synthetic waveforms and observation, especially for surface waves, may reflect the discrepancies of three-dimensional mantle structure used in this simulation with the actual Earth.

1 2011 Tohoku Earthquake

The 2011 Tohoku earthquake (March 11, 2011, 38.322N 142.369E, depth 24 km Mw9.0 by JMA) was one of the largest earthquakes recorded during the past 100 years all over the globe. The earthquake was felt in large parts of the Japanese Islands and the huge tsunami caused serious damages to the Pacific coasts of

S. Tsuboi (✉) · T. Nakamura
JAMSTEC, Yokohama 236-0001 Japan
e-mail: tsuboi@jamstec.go.jp; t_nakamura@jamstec.go.jp

M. Resch et al. (eds.), *Sustained Simulation Performance 2012*,
DOI 10.1007/978-3-642-32454-3_9,

northern part of Japanese Islands. Although the epicenter of this earthquake locates at the place of subduction of oceanic tectonic plate, where we expect to have frequent M7 or 8 class large earthquakes, there was no historical record of this size of huge earthquake in this area. This fact raises a question if there are different characteristics of the rupture mechanism of this earthquake, which marks this earthquake as a peculiar event and many researches are conducted to reveal source characteristics of this earthquake.

2 Earthquake Rupture Mechanism

We apply the waveform inversion [1, 2] to obtain slip distribution in the source fault at the 2011 Tohoku earthquake in the same manner as our previous work [3]. We use 22 broadband seismograms of IRIS GSN seismic stations with epicentral distance between 30 and 100 degrees. The broadband original data are integrated into ground displacement and band-pass filtered in the frequency band 0.002–1 Hz. We use the velocity structure model IASP91 [4] to calculate the wavefield near source and stations. We assume that the strike of the fault plane is 201 degree and the dip angle is 9 degree, based on Global Centroid Moment Tensor solution. The length of subfault used for our inversion is 30 km in the strike direction and 20 km in the dip direction. The assumed fault length is totally 460 km consistent with the aftershock distribution. The nonnegative least-squares method [5] is employed for constraining the rake angle in the waveform inversion. The strike and the total length of the source fault are illustrated in Fig. 1 with location of epicenter. We assume that rupture velocity along the fault is about 2.0 km/s. The results of the inversion show the bilateral rupture to the northeast and southwest with two main asperities along the fault; the maximum slip of 49 m with the reverse fault mechanism at approximately 100 km northeast of the epicenter and another large slip with reverse fault mechanism at 100 km southwest of the epicenter. The total amount of the released seismic moment corresponds to moment magnitude Mw = 9.1. The slip distribution along the fault surface is also shown in Fig. 1. The duration of source time function is 150 s. The fault length of 450 km and the source duration time of 150 s are typical for Mw 9.1 earthquake. However, the maximum slip of 49 m is unusually large. Also the rupture velocity of 2.0 km/s may characterize this earthquake as slow tsunami generating earthquake. We should note that the fault rupture model, we have obtained, is not unique in the sense that the variance reduction is not significant. There are possibilities of other models, which show much slower rupture velocity or larger slip, although this result is basically consistent to other rupture models [6, 7]. Thus, we examine the validity of these fault rupture parameters by comparing the theoretical seismograms computed for this fault model with the observed seismograms.

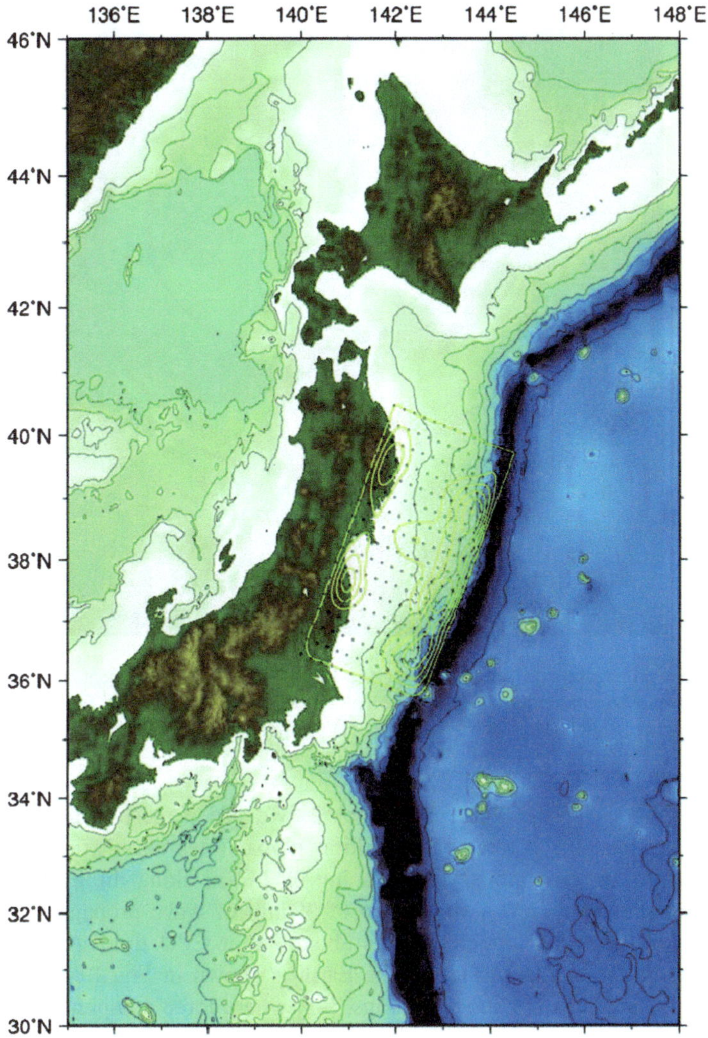

Fig. 1 Slip distribution projected on the surface topography and bathymetry. Slip distribution along the fault surface, of which strike of the fault plane is 201 degree, measured clockwise from north and the dip angle is 9 degree, measured downward from the horizontal plane, is shown as contour map

3 Broadband Synthetic Seismograms

We calculate broadband synthetic seismograms with this source propagation model for a realistic 3D Earth model using the spectral-element method [8, 9]. We use the Earth Simulator 2 of JAMSTEC to compute synthetic seismograms using the spectral-element method. The simulations are performed on 726 processors, which

Fig. 2 Wave propagation at the surface of the Earth. Vertical displacement at the surface is exaggerated and large amplitude regions are colored in *red*. Movie created from this figures is shown at following URL (http://www.youtube.com/user/jamstecchannel#p/a/u/0/j4tLFeJiAhY)

require 91 nodes of the Earth Simulator 2. We use a mesh with 200 million spectral-elements, for a total of 13 billion global integration grid points. This translates into an approximate grid spacing of 2.0 km along the Earth's surface. On this number of nodes, a simulation of 30 min of wave propagation accurate at periods of 3.5 s and longer requires about 7 h of CPU time. An example of wave propagation at the surface of the Earth is shown in Fig. 2. Also examples of waveform matches are shown in Figs. 3–5, where three component broadband seismograms are compared with the observations at teleseismic stations, AFI (Afiamalu, Samoa Islands, epicentral distance 67 degrees), ESK (Edinburgh, UK, 81 degrees), and HRV (Harvard, USA, 92 degrees). Each trace is bandpass filtered between 0.002 and 0.1 Hz. We may say that the synthetic seismograms reproduces observed seismograms generally well for body waves. Although surface waves are not modeled well, but it is because the surface waves with this period range depend on shallow crustal structure, which may not be modeled well in the 3D mantle model. Figure 6 demonstrates example of comparison in lower frequency range for station AFI and shows that the agreement is excellent in lower frequency range.

Figure 7 shows synthetics for near field station KSK (3.1 degree), of F-net broadband network in the coast of Tohoku region. Because of proximity of the station location to the epicenter, the displacement shows static displacement, whose

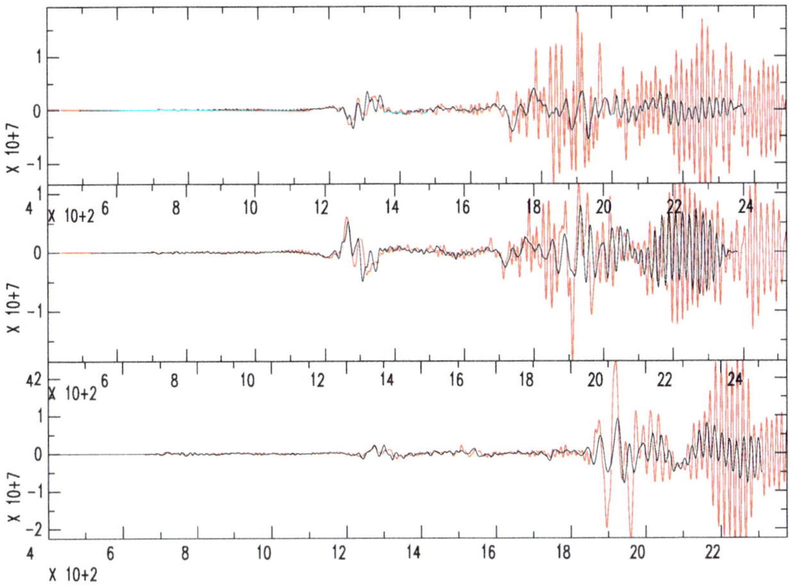

Fig. 3 Comparisons of synthetic seismograms and observation for IRIS GSN station AFI. The synthetics and the observations are in *red* and *black*, respectively. Instrument responses are convolved to synthetics to convert them to ground velocity. Traces are EW, NS, and UD components from *top* to *bottom*, respectively. The origin of the time axis is origin time of the event and the vertical axis shows digital count. All of the traces are bandpass filtered between 500 s and 10 s

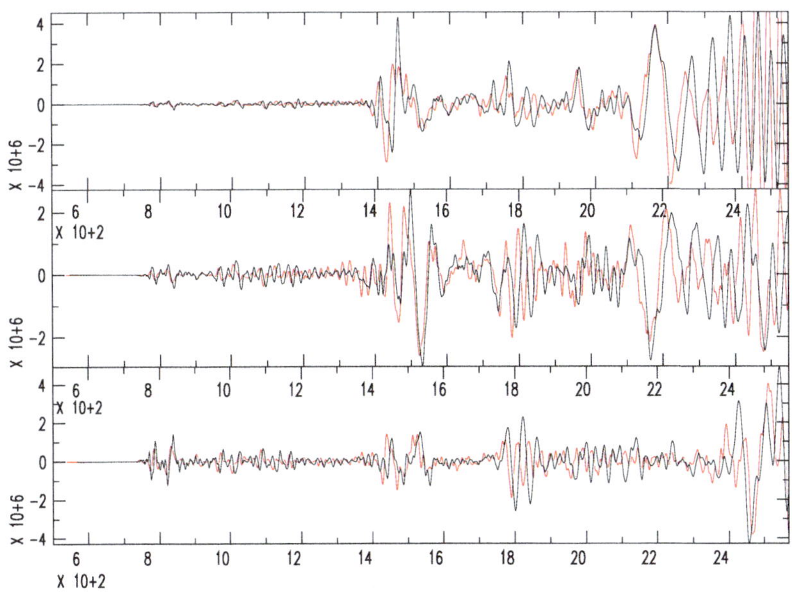

Fig. 4 Same as Fig. 3 but for station ESK

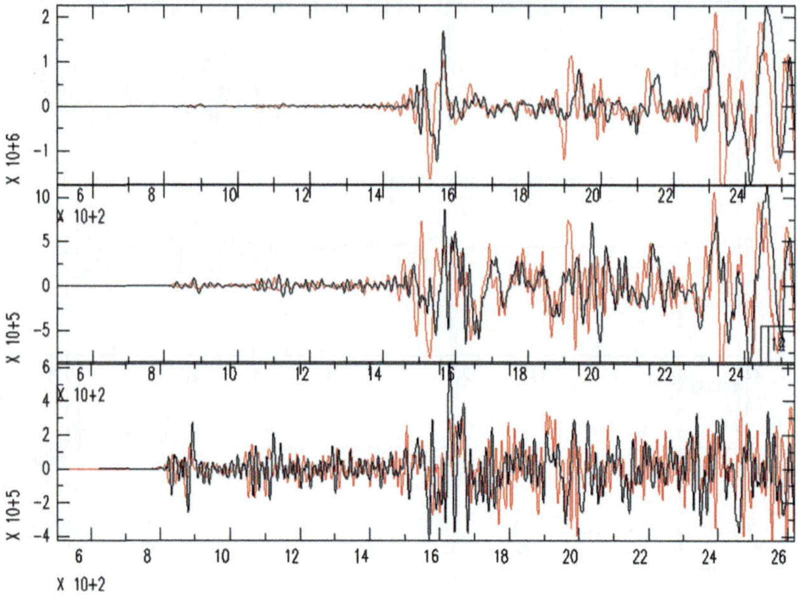

Fig. 5 Same as Fig. 3 but for station HRV

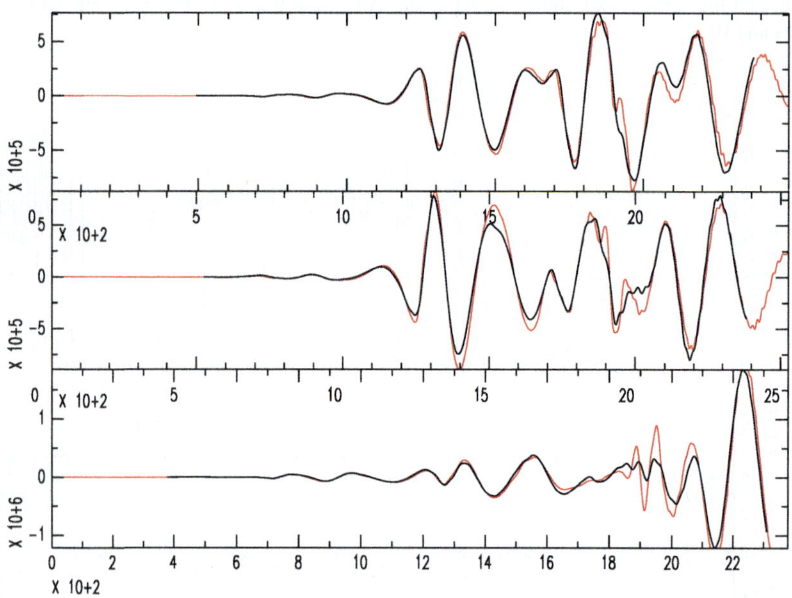

Fig. 6 Comparisons of synthetic seismograms for station AFI. Bandpass filtered at 500 s and 200 s

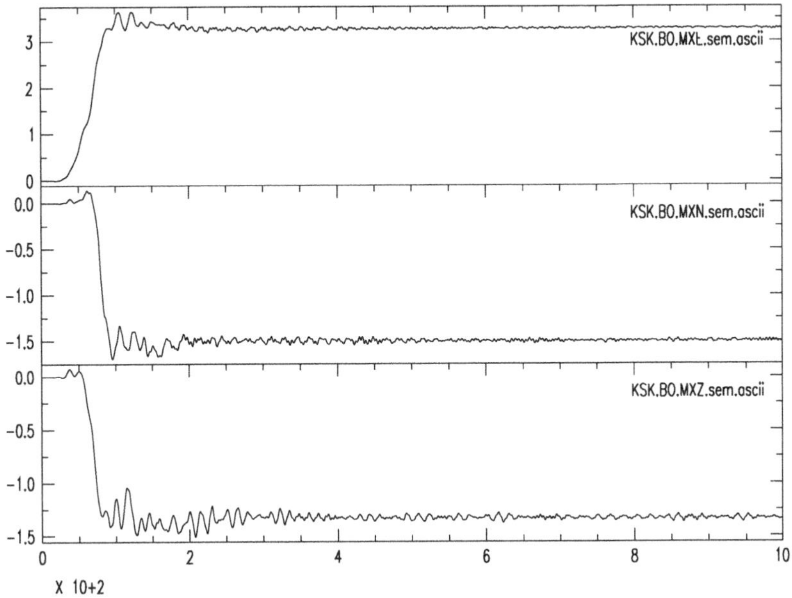

Fig. 7 Synthetic seismograms of Fnet network stations, KSK. Three component theoretical displacement siesmograms are shown. 1,000 s records of EW,NS, UD are shown from *top* to *bottom*. Vertical unit is in meters. Estimated static displacements are east 3.2 m, south 1.5 m, and down 1.3 m for KSK

pattern is consistent to the observed crustal deformation [10]. These results indicate that the synthetic seismograms computed for this finite source model reproduce observed seismograms for both near field and teleseismic stations, which suggest that this source model captures rupture properties of this earthquake.

References

1. M. Kikuchi, H. Kanamori, Inversion of complex body waves. III. Bull. Seismol. Soc. Am. 81, 2335 (1991).
2. M. Kikuchi, H. Kanamori, Note on Teleseismic Body-Wave Inversion Program. http://www.eri.u-tokyo.ac.jp/ETAL/KIKUCHI/ (2003).
3. T. Nakamura S. Tsuboi, Y. Kaneda, Y. Yamanaka, Rupture process of the 2008 Wenchuan, China earthquake inferred from teleseismic waveform inversion and forward modeling of broadband seismic waves. Tectonophysics. 491, 72 (2010).
4. B. L. N. Kennet, E. R. Engdahl, Traveltimes for global earthquake location and phase identification. Geophys. J. Int. 105, 429 (1991).
5. C. L. Lawson, R. J. Hanson, Solving Least Squares Problems, Prentice-Hall, New Jersey, (1974)
6. K. Koketsu et al., A unified source model for the 2011 Tohoku earthquake. Earth Planet. Sci. Lett. 310, 480 (2011).

7. S. J. Lee, B. S. Huang, M. Ando, H. C. Chiu, J. H. Wang, Evidence of large scale repeating slip during the 2011 Tohoku-Oki earthquake. Geophys. Res. Lett. 38, L19306 (2011).
8. D. Komatitsch, J. Ritsema, J. Tromp, The spectral-element method, Beowulf computing, and global seismology. Science 298, 1737 (2002).
9. S. Tsuboi, D. Komatitsch, C. Ji, J. Tromp, Broadband modelling of the 2002 Denali fault earthquake on the Earth Simulator. Phys. Earth Planet. Inter. 139, 305 (2003).
10. S. Ozawa et al., Coseismic and postseismic slip of the 2011 magnitude-9 Tohoku-Oki earthquake. Nature 475, 373 (2011).

Part IV
Computational Engineering Applications and Coupled Multi-physics Simulations

A Framework for the Numerical Simulation of Early Stage Aneurysm Development with the Lattice Boltzmann Method

J. Bernsdorf, J. Qi, H. Klimach, and S. Roller

Abstract In this paper, we describe a new approach towards numerical simulation of flow induced early stage development of cerebral aneurysm. The wall shear stress gradient, computed by a CFD simulation inside a bifurcating flow channel, triggers a physiological process leading to the remodelling, and in the worst case, degeneration of the vessel walls. The lattice Boltzmann method, extended by a generic vessel wall model to allow an efficient modification of the flow geometry during run-time, is employed for simulating the modification of the vessel wall, which is considered as initial step for aneurysm formation. First results presented here show a thinning of the vessel wall at locations left and right of the apex of the bifurcation, in good agreement with experimental studies.

1 Introduction

Aneurysms are extreme widenings of vessels which can be, if they rupture, life threatening. For fully developed cerebral aneurysms (CA), the estimation of the rupture risk has been studied within various research projects [1], and simulation supported treatment planning is subject to ongoing research [2, 3].

In contrast to this, the process of early stage development of CA is not yet well understood. The locations of typical CAs at proximal arterial bifurcations and along the outer curvatures of intracranial vessels implicate a significant contribution of haemodynamic parameters for the formation process [4]. Experimental studies indicate that remodelling of the artery is a physiological reaction on a combination of high wall shear stress (WSS) and high WSS gradient (WSSG) near the apex of a carotid bifurcation [5].

J. Bernsdorf (✉) · J. Qi · H. Klimach · S. Roller
German Research School for Simulation Sciences GmbH, Schinkelstr. 2a, 52062 Aachen, Germany and RWTH Aachen University, Templergraben 55 52056 Aachen, Germany
e-mail: j.bernsdorf@grs-sim.de; j.qi@grs-sim.de; h.klimach@grs-sim.de; s.roller@grs-sim.de

M. Resch et al. (eds.), *Sustained Simulation Performance 2012*,
DOI 10.1007/978-3-642-32454-3_10,

In this paper, we suggest a novel approach towards the lattice Boltzmann based simulation of flow induced early stage formation of cerebral aneurysm. We show that the combination of a lattice Boltzmann flow solver and a simple generic vessel wall model, reacting only to flow properties, leads to modification of the vessel walls at experimentally predicted locations. We further demonstrate the ability and performance of the underlying simulation package, which forms the basis for a later extension towards improved models coupling the output of the flow solver with a more detailed and accurate biological model for the vessel wall.

The next sections are organised in the following way: first, we introduce the medical problem and biological aspects of flow-induced remodelling of cerebral arteries. Then, our approach of extending a lattice Boltzmann flow solver for simulating this process is described and we discuss performance considerations. Preliminary results are presented in Sect. 5, followed by an outlook describing our strategy for further extending the method.

2 Medical Problem and Biological Process

Simply speaking, the initial process of aneurysm development can be summarised as follows: the innermost cellular layer of an artery (the endothelial cells) is able to react to the shear rate of the flow. Changed flow conditions trigger a remodelling of the vessel wall (as described in more detail below), which is an adaptive process to restore required flow rates for keeping up the functionality of the arterial network. Aneurysm development can in that way be understood as an adoption to changed flow conditions.

Modifications in the arterial network, caused either by unhealthy life-style (leading to arteriosclerosis) or surgery (e.g., by-pass or stenting), can dramatically change the distal flow properties (in the network "upstream" of the affected area). When this results in higher flow rates (e.g., through re-opened blockages or by compensating occlusion of a parallel flow channel), the affected arteries remodel in an attempt to adapt to the new flow properties. Thus, the flow provides conditions for a remodelling of the vessel wall and thinning of the internal lamina. Eventually, this degenerative biological process can initiate the formation of an aneurysm.

Since the key players for the remodelling, namely a high wall shear stress gradient, are known from animal experiments, numerical parameter studies to identify typical configurations of geometry and flow properties leading to aneurysm formation become a possibility. Although biological details are not yet fully clear, we can already figure out the general chain of some critical factors in this remodelling process, leading to aneurysm initiation.

High wall shear stress (WSS) and WSS gradient, which occur through changes in the arterial network, cause dysfunction of endothelial cells by migrating along the accelerating flow, detaching from each other [6], or even worse, getting damaged [7]. The smooth muscle cells, which lay next to the endothelium, are then exposed to these frictional forces and degraded as the endothelial cells. The dysfunction and the decrease in the density of both endothelial cells and smooth muscle cells further lead to remodelling of the extracellular matrix, which is essential for structural maintenance of the vessel wall [8]. Due to that process, the blood vessel loses its mechanical support, and might eventually suffer local expansion through hydrostatic pressure. Therefore, this local de-stabilisation of the vessel wall can be considered as the first step in the development of an aneurysm.

A better understanding of why and where cerebral aneurysm (CA) develop can significantly contribute to fundamental research, as well as patient-specific treatment planning. As a long-term perspective, we can imagine the patient-specific estimation of the risk to develop CA, ideally prior to a planned treatment such as stenting or by-pass operation.

3 Simulation Approach

A complete model for the numerical simulation of flow-induced thinning of the vessel wall, as described in the previous paragraph, must fulfil the following requirements:

- Flow properties (as the wall shear stress and its gradient) must be efficiently and locally determined in complex changing geometries.
- A significant modification of the flow domain, based on computed flow properties, must be possible during run-time of the simulation.
- A numerical model simulating the previously described biological reaction of endothelial and smooth muscle cells to the flow must be developed.
- The various time-scales for the process (below seconds for the flow and above days for the aneurysm formation) have to be considered within a multi-scale simulation approach [9].

The lattice Boltzmann (LB) method allows an efficient local computation of the shear stress from the non-equilibrium part of the density distributions in non-Newtonian flow through complex aneurysm geometries [10]. Also, changing solid boundary conditions during run-time have been successfully applied in the context of medical-physics simulations [11].

Our suggested extension of a standard LB flow solver consists of iterating the following steps within the framework of a coupled simulation:

- Compute the relevant flow parameters for the current geometry, namely WSS and WSS gradient.
- Communicate these data to the biological model.
- Compute the flow-induced reformation of the cellular topology (in this paper achieved by a first simplified approach).
- Communicate the increment of the geometric boundaries to the flow solver.

For the preliminary studies presented in this paper, we reduced the complexity of the simulation approach by collapsing all biological processes into a WSSG threshold model: above a certain WSSG value, the loss of internal elastic lamina is modelled by turning a solid lattice node into a fluid node. This first simple approach can already indicate if a remodelling of the arterial wall occurs in the expected area.

4 Performance Considerations

As explained in the previous sections, the modelling of the overall process is fairly complex and requires the interaction of flow and physiology. Further the flow simulation requires a relatively high resolution to accurately describe the small scale boundary layer properties. Therefore, it is essential to use a highly efficient implementation for the flow simulation, while maintaining the flexibility for interactions with the additional simulation influencing factors.

The lattice Boltzmann method in itself is very well suited for efficient computations, due to the very compact kernel, which can be highly optimised. In our solver we use an octree data structure to represent the mesh, which allows fast access and change to geometrical properties (see e.g., [12]). At the same time the computational kernel is kept in a uniform mesh, where all data is explicitly accessible as in unstructured meshes. None of the logic for the various tasks in the embedding framework has to be put into the kernel. This enables a high degree of optimisations for the kernel itself, while maintaining the flexibility to change the mesh and represent arbitrary complex geometries. Furthermore the octree data structure describing the mesh also allows for an efficient description of the partitions in a parallel computation. With the inherent knowledge about the topology of the tree it is possible to locally compute neighbour relations on each partitions and decide neighbourhood relations without larger amounts of communications. Information can be kept local to a large degree, which is very important for transient simulations as described in this work with geometric changes.

The approach was successfully shown to scale to hundred thousand processes and thus allow full usage of todays supercomputing facilities. Most of the running time is spent in the kernel, which does not notice the outer complexity of the embedding program. For this reason a high sustained performance of around 10 % of the theoretical peak performance is achieved by the overall application in serial. As the mesh organisation is completely unstructured, this performance is not influenced by the complexity of the geometry. In fact in parallel runs, the simulation might even be faster for complex geometries, due to the reduced communication surfaces by the introduced walls, which decouple the partitions in the fragmented computational domain. This is shown in the speed-up comparison between a generic cube with 134 million elements and no boundaries at all and a rather complex geometry resembling a porous media with 66 million elements in Fig. 1. The graph shows the performance in million lattice updates per second over the number of processes on the Cray XE 6 system Hermit at the HLRS in Stuttgart. As can be seen, the absolute performance is not affected by the complex geometry on small process counts and the strong scaling is even improved by the walls subdividing the overall domain.

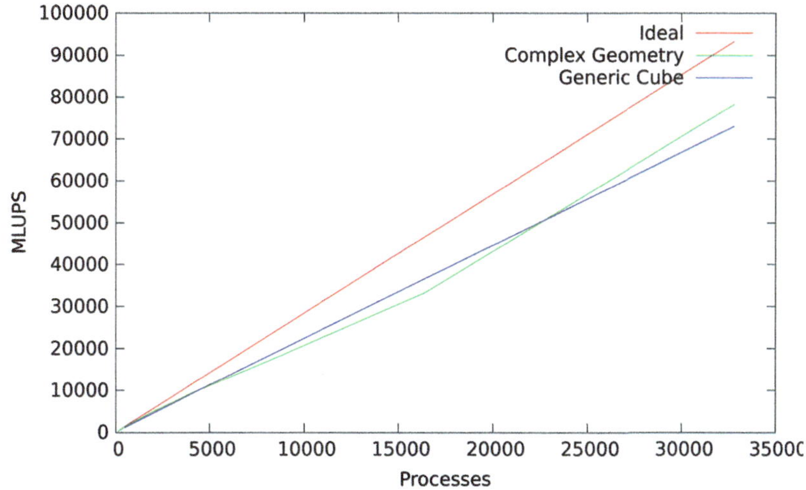

Fig. 1 Speed up comparison on Hermit between domains with geometry and without

5 Results

5.1 Simulation Setup

A lattice BGK implementation was extended to compute the wall shear stress from the non-equilibrium part of the density distribution function, and from that to compute a local wall shear stress gradient (WSSG). Further, the conversion of a solid node into a fluid node was enabled, if at least one fluid node adjacent to the solid node under consideration a certain threshold of the WSSG was reached. This simple approach mimics the thinning of the vessel wall, it will be replaced by a more complex model taking into account details of the biological process in later steps.

New fluid nodes were primed by an equilibrium distribution for zero flow velocity. To ensure that no artefacts from initialisation of the fluid nodes will disturb the WSS computation and further progress of the simulation, the flow field was allowed to adopt to the new configuration with a sufficient number (order of several hundred in our case) of iterations, before the WSS measurement was re-activated.

A simple initial 2-D flow geometry modelling a symmetric bifurcation was constructed (Fig. 2), with velocity inlet and pressure outlet boundaries. At the walls, half way bounce back boundary condition was employed.

The region of interest (shown in Fig. 3) had a resolution of $lx \times ly = 100 \times 100$ lattice nodes. After reaching a steady state flow profile after approximately 4,000 iterations, the functionality for measuring the wall shear stress gradient and for

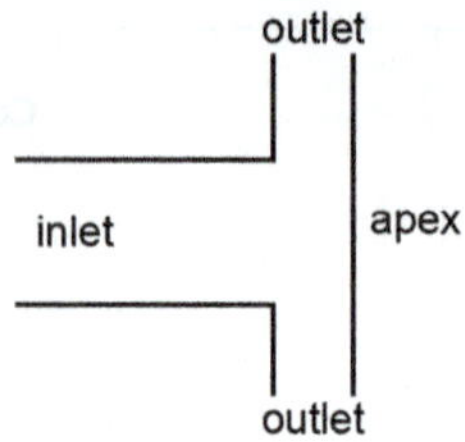

Fig. 2 Initial flow geometry, inlet, outlets and the apex are indicated

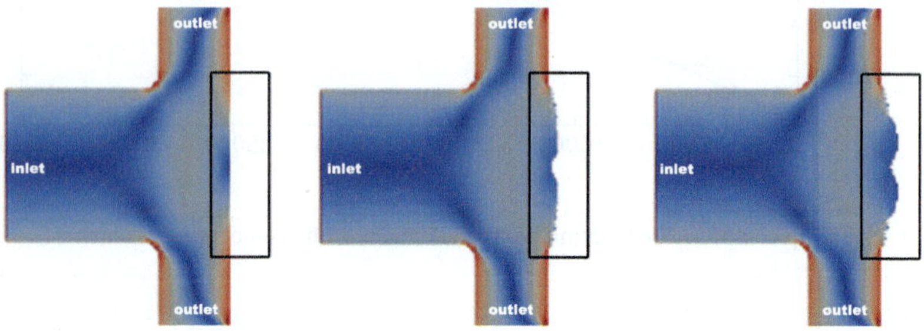

Fig. 3 Thinning of the vessel wall, time increasing from *left* (initial geometry) to *right* (steady state). The *colour* indicates the wall shear stress (WSS, *red* = high)

turning solid into fluid cells was activated. In this preliminary study, the WSSG threshold was set to a sufficient value to produce an effect, and not yet directly correlated to data from experimental data.

5.2 *Observations*

Near the apex of the bifurcation, a sharp increase of the WSS can be observed (see Fig. 3). As expected (and in good agreement with numerical simulations reported in [4]) the maximum of the WSSG can be observed left and right upstream of the apex.

After turning on the subroutine modelling the biological process, a reduction of solid cells (representing the vessel wall) can be observed at the regions of maximum WSSG (see Fig. 3).

This reduction of solid cells is interpreted in our model as loss of internal elastic lamina, or thinning of the vessel wall, which later would lead to the formation of an aneurysm.

After a certain time, the adaptation process comes to an end, and a steady state is reached.

5.3 *Discussion*

The resulting reduction of solid cells left and right downstream the apex of the bifurcation correlates with the experimentally observed thinning of the vessel walls, and the location is qualitatively in good agreement with data reported in the literature (see e.g., [4]). Our much simplified initial approach of simulating the first step in a flow induced aneurysm growth can therefore be considered as qualitatively successful, in spite of the simplicity of the model employed.

6 Conclusion and Outlook

We presented a new model for the numerical simulation of flow induced early stage cerebral aneurysm development. First simulation results, produced with a simplified approach based on an extended 3-D lattice Boltzmann solver, show results in qualitatively good agreement with experimental data.

In the future, this approach will be extended in various ways to produce quantitatively correct results in patient specific geometries:

- The biological routines, currently only acting on one threshold parameter to turn solid cells into fluids, will significantly be refined and extended, to take into account relevant biological processes for a correct simulation of the cell migration.
- With this extended biological model, it might become necessary to integrate the lattice Boltzmann (LB) flow solver into a multi-scale simulation environment, to take into account the different time-scales of biological and flow processes.
- A three-dimensional flow geometry will be employed, and the half way bounce-back wall boundary condition will be replaced by a more sophisticated model, to allow an accurate computation of the wall shear stress.
- Transient inlet- and outlet boundary conditions will be employed for a physiological flow velocity and pressure of the cardiac cycle. This might require coupling the flow solver to a systemic model.
- A structure model has to be coupled, in order to simulate the actual development of the aneurysm shape, caused by a mechanical reaction of the thinning vessel wall to the flow.

This complete model will be a coupled simulation of transient fluid flow in changing geometries, a biological model for flow induced degeneration of arterial cells, and a structure mechanics model to compute the aneurysm shape from flow properties and mechanical properties of the degenerating vessel wall. Performance and efficiency of the solver are necessary prerequisites for simulating the complex coupled biology and flow processes. In this paper we demonstrated the applicability of the underlying approach and the suitability of the flow solver and its extensions.

References

1. see e.g., the aneurIST IST project: http://www.aneurist.org
2. see e.g., the Thrombus ICT project: http://www.thrombus-vph.eu
3. J. Bernsdorf, D. Wang and G. Berti. Two Complementary Approaches for Integrating a Lattice Boltzmann Flow Solver into Simulation Frameworks. Proceedings of the International Conference on Computational Science, ICCS 2011. Procedia Computer Science Volume 4 (2011): 1014–1020
4. E. Metaxa, M. Tremmel, S.K. Natarajan, J. Xiang, R.A. Paluch, M. Mandelbaum, A.H. Siddiqui, J. Kolega, J.M. and H. Meng. Characterization of Critical Hemodynamics Contributing to Aneurysmal Remodeling at the Basilar Terminus in a Rabbit Model. Stroke 41 (2010): 1774–1782
5. Z. Wang, J. Kolega, Y. Hoi, L. Gao, D. Swartz, E.I. Levy, J. Mocco and H. Meng. Molecular Alterations Associated with Aneurysmal Remodeling are Localized in the High Hemodynamic Stress Region of a Created Carotid Bifurcation. Neurosurgery 65(1) (2009): 169–178
6. M.P. Szymanski, E. Metaxa, H. Meng, J. Kolega. Endothelial cell layer subjected to impinging flow mimicking the apex of an arterial bifurcation. Annals of biomedical engineering 36(10) (2008): 1681–1689
7. H. Meng, D. D. Swartz, Z. Wang, Y. Hoi, J. Kolega, E. M. Metaxa, M. P. Szymanski, J. Yamamoto, E. Sauvageau, and E. I. Levy. A Model System for Mapping Vascular Responses to Complex Hemodynamics at Arterial Bifurcations In Vivo. Neurosurgery 59(5) (2006):1094–1101
8. D. Krex, H. K. Schackert and G. Schackert. Genesis of Cerebral Aneurysms - An Update. Acta Neurochir (Wien) 143(5) (2001): 429–448
9. A. Caiazzo, D. Evans, J.-L. Falcone, J. Hegewald, E. Lorenz, B. Stahl, D.Wang, J.Bernsdorf, B. Chopard, J. Gunn, R. Hose, M. Krafczyk, P. Lawford, R. Smallwood, D. Walker, A. Hoekstra. A Complex Automata approach for in-stent restenosis: Two-dimensional multiscale modelling and simulations. Journal of Computational Science, Volume 2, Issue 1(2011): 9–17
10. J. Bernsdorf and D. Wang. Non-Newtonian blood flow simulation in cerebral aneurysms. Computers & Mathematics with Applications, Volume 58, Issue 5 (2009): 1024–1029
11. S.E. Harrison, S.M. Smith, J. Bernsdorf, D.R. Hose, P.V. Lawford. Application and validation of the lattice Boltzmann method for modelling flow-related clotting. Journal of Biomechanics, Volume 40, Issue 13 (2007): 3023–3028
12. S. Roller, J. Bernsdorf, H. Klimach, M. Hasert, D. Harlacher, M. Cakircali, S. Zimny, K. Masilamani, L. Didinger, J. Zudro. An Adaptable Simulation Framework Based on a Linearized Octree. In: M. Resch et al. (eds.) High Performance Computing on Vector Systems 2011, springer (2012): 93–105

Performance Evaluation of a Next-Generation CFD on Various Supercomputing Systems

Kazuhiko Komatsu, Takashi Soga, Ryusuke Egawa, Hiroyuki Takizawa, and Hiroaki Kobayashi

Abstract The Building-Cube Method (BCM) has been proposed as a new CFD method for an efficient three-dimensional flow simulation on large-scale supercomputing systems, and is based on equally-spaced Cartesian meshes. As a flow domain can be divided into equally-partitioned cells due to the equally-spaced meshes, the flow computations can be divided to partial computations of the same computational cost. To achieve a high sustained performance, architecture-aware implementations and optimizations considering characteristics of supercomputing systems are essential because there have been various types of supercomputing systems such as a scalar type, a vector type, and an accelerator type. This paper discusses the architecture-aware implementations and optimizations for various supercomputing systems such as an Intel Nehalem-EP cluster, an Intel Nehalem-EX cluster, Fujitsu FX-1, Hitachi SR16000 M1, NEC SX-9, and a GPU cluster, and analyses their sustained performance for BCM. The performance analysis shows that memory and network capabilities largely affect the performance of BCM rather than computational potentials.

K. Komatsu (✉) · R. Egawa · H. Kobayashi
Cyberscience Center, Tohoku University/JST CREST, 6–3 Aramaki-aza-aoba, Aoba, Sendai 980–8578, Japan
e-mail: komatsu@isc.tohoku.ac.jp; egawa@isc.tohoku.ac.jp; koba@isc.tohoku.ac.jp

T. Soga
NEC System Technologies, Ltd., Osaka 540–8551, Japan
e-mail: soga-txa@necst.nec.co.jp

H. Takizawa
Graduate School of Information Sciences, Tohoku University/JST CREST, 6-6-01 Aramaki-aza-aoba, Aoba, Sendai 980–8579, Japan
e-mail: tacky@isc.tohoku.ac.jp

M. Resch et al. (eds.), *Sustained Simulation Performance 2012*,
DOI 10.1007/978-3-642-32454-3_11, © Springer-Verlag Berlin Heidelberg 2013

1 Introduction

Since 1960s, the numerical calculation using computers has been utilized for simulations and analysis of fluid dynamics. In CFD, unstructured mesh and boundary-fitted mesh have generally been utilized to represent complicated geometries such as a three-dimensional whole airplane. These mesh methods have advantages of the mesh quality and the accuracy of the simulations. However, CFD algorithms with these mesh methods become complicated, and need a high computational cost. In addition, because the mesh is not regular, it is difficult to realize a spatial higher-order flow solver.

In order to solve these problems, the *Building-Cube Method (BCM)* has been proposed to efficiently simulate various fluids [4, 7, 8]. BCM uses equally-spaced Cartesian meshes. One of the advantages of BCM is that the algorithms of pre-processing, post-processing, and even the flow solver can be simplified [2] because of the equally-spaced Cartesian meshes. Another advantage is that it is well suited for highly parallel computation because equally-spaced meshes produce many parallel tasks with the same amount of computation.

Along with the development of CFD, the performance of supercomputing systems has also drastically been improved because of the rapid advancement of semiconductor technologies. To achieve a high sustained performance using supercomputing systems, architecture-aware implementation and optimization considering characteristics of supercomputing systems are essential due to various types of supercomputing systems such as a scalar type, a vector type, an accelerator type.

This paper describes architecture-aware implementations and optimizations of BCM on an Intel Nehalem-EP cluster, an Intel Nehalem-EX cluster, Fujitsu FX-1, Hitachi SR16000 M1, NEC SX-9, and a GPU cluster to examine the implication of their architectural features with BCM. From experimental results, this paper analyses the performances and scalabilities of BCM.

2 Overview of the Building Cube Method

BCM is designed for three-dimensional large-scale flow computations around practical geometries using high-density grids [4]. The basic idea of BCM is to decompose a whole flow domain into sub-domains called *cubes*, and further decompose each cube into high-density and equally-spaced Cartesian meshes called *cells* shown in Fig. 1. The size of each cube is determined by geometries and flow features at its location [2].

One of the advantages of BCM is that the algorithm is simple because it does not deal with complicated mesh structures. Thus, the simplicity of pre-processing, flow solvers, and post-processing lead to fast computation.

Another advantage of BCM is that the calculations of cubes can easily be decomposed into many data parallel tasks of the same size because the calculations

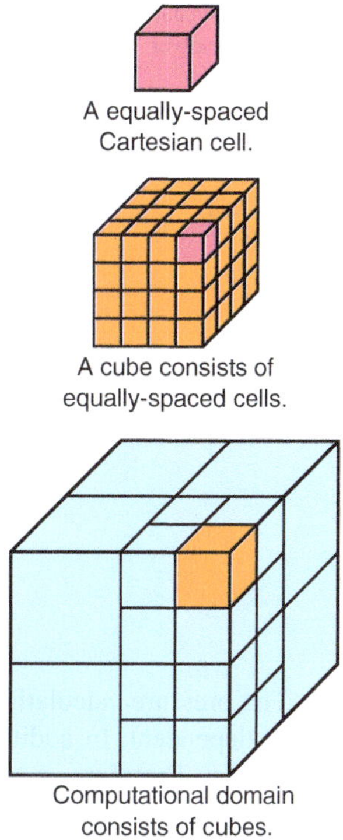

Fig. 1 Computational mesh in BCM

are independent each other. In addition, each cube has the same computational cost and data size for the calculations. Although there are data dependencies among adjacent cells in a cube, the computational cost per cell is also the same. More massive data parallelism in BCM might be obtained if the dependencies are eliminated.

The flowchart of the BCM incompressible flow solver is shown in Fig. 2 [7, 8]. The governing equations are incompressible Navier–Stokes equations. The fractional-step method [1, 3, 6] is used with the finite difference scheme on the staggered arrangement. In the fractional-step method, the solver can be classified into three major stages in one time step; a solver stage for calculating a temporal velocity field, a solver stage for calculating a pressure field, and a solver stage for calculating a real velocity field. In each stage, calculations of its field and data exchanges between cubes are included. The most dominant part in these stages is the calculation of the pressure field by solving the Poisson equation using the SOR method.

To calculate the pressure of one cell, a seven-point stencil calculation, which requires the pressure data of a cell and its six adjacent cells, is performed. As the stencil calculations for all cells in all cubes are repeated until the difference of the calculated field is convergent, the calculations of the pressure field dominates the most of time for the whole BCM calculations.

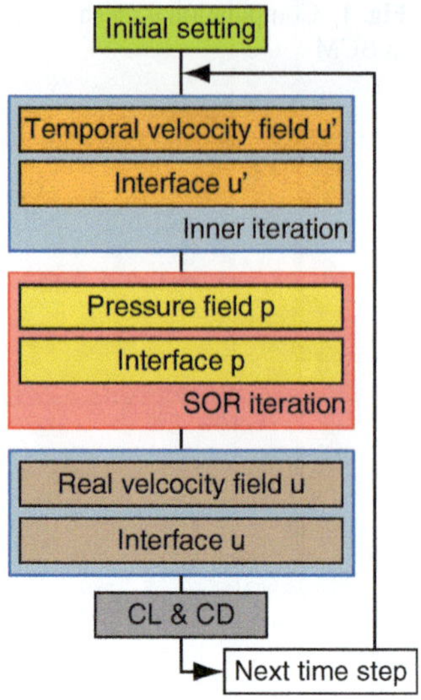

Fig. 2 Flowchart of the BCM flow solver

The pressure calculations for cubes can be performed in parallel because they are independent. In addition, the computational cost of each cube is completely the same. Therefore, parallel computing is adequate to accelerate the pressure calculations.

3 Implementation of BCM on Various Systems

To achieve significant acceleration by parallel processing on supercomputing systems, architecture-aware implementations and optimizations considering characteristics of supercomputers are essential.

This section describes the overview of the target supercomputing systems. The specifications of processors used in the systems are shown in Table 1. Then, the architecture-aware implementations and optimizations of BCM for these supercomputers are described as follows.

3.1 Implementation on Scalar Systems

The Nehalem-EP cluster, the Nehalem-EX cluster, FX-1, and SR16000 M1 are scalar parallel supercomputers that equip Nehalem-EP, Nehalem-EX, SPARC64VII, and Power7 processors, respectively. As shown in Table 1, these scalar processors also have large on-chip cache memories. On-chip L2 and/or L3 caches should be used for data with high locality to avoid redundant memory accesses. Moreover, uses of SIMD instructions are essential to efficiently process multiple data.

Table 1 Specifications of a processor in the supercomputing systems

System	GFlops/s	Mem.BW (GB/s)	# of Cores	On-chip memory	B/F
Nehalem EP	46.93	25.6	4	256 KB L2/core, 8 MB shared L3	0.55
Nehalem EX	74.48	34.1	8	256 KB L2/core, 24 MB shared L3	0.47
SPARC64VII	40.32	40.0	4	6 MB shared L2	1.0
Power 7	245.1	128	8	256 KB L2/core, 32 MB shared L3	0.52
SX-9	102.4	256	1	256 KB ADB	2.5
Tesla C1060	78	102	1	16 KB/SM	1.3

Although the Red–Black method can eliminate the data dependency, the stride memory accesses are required, resulting in performance degradation. Even though dividing an array into two arrays, non-unit-stride accesses are required. Thus, in the implementation on the scalar systems, the original SOR method is adopted to avoid degrading the performance by the stride memory accesses.

In the implementation of BCM on a scalar system, cubes are hierarchically assigned to nodes and then to processors in a node. As the computational cost of each cube is also the same, efficient parallel processing using a number of scalar processors can be carried out.

3.2 *Implementation on a Vector System*

SX-9 is a vector parallel supercomputer consisting of a large Symmetric Multi-Processing (SMP) nodes, each of which has 16 102.4Gflop/s-vector processors. In the implementation on SX-9, the Red–Black SOR method using mask tables is used. The Red–Black method can avoid indirect memory accesses and exploit the data parallelism among cells by removing the dependencies among cells. As parallelizing the SOR method generally shortens the length of loop, the mask tables can avoid accessing unnecessary data without shortening the length of a loop. Thus, the loop remains long enough to utilize all of the vector units of SX-9.

The effective use of an on-chip 256KB software-controllable cache named Assignable Data Buffer (ADB) in SX-9 is also a key to exploit the potential of SX-9. Once data specified by programmers are accessed, these data are stored in ADB and can be used in the next accesses. Thus, the ON_ADB directives are inserted to the source code to specify reusable data in the seven-point stencil calculations. As a result, the reusable data are kept in ADB at runtime. As the main memory and ADB can simultaneously provide data to vector pipelines, the vector processor can access those data at a high sustained bandwidth, and thereby achieve a high performance.

3.3 Implementation on a GPU System

A GPU cluster consists of multiple nodes, each of which has a CPU with main memory and one or more GPUs. Each GPU can be considered as a many-core processor consisting of hundreds of *stream processors (SPs)* in *CUDA (Compute Unified Device Architecture)* [5]. In CUDA, SPs are grouped into *stream multiprocessors (SMs)*, and several SPs in an SM work together.

The memory system of the CUDA platform is hierarchical. The *shared memory* is an on-chip memory space shared by threads. The capacity of the shared memory is small, but the memory access latency is very short. On the other hand, the *global memory* is the largest off-chip memory, but it needs a long access latency. Thus, it is necessary to use both shared memory and global memory appropriately. Key techniques for efficient data transfers are to make good use of the shared memory and to use coalesced global memory accesses as much as possible.

Massive parallelism of BCM is well suited for load balancing among nodes and also among many cores in each GPU. In the implementation of BCM on a GPU cluster system, after grouping cubes into subsets, each subset of cubes is assigned to one of GPU nodes as shown in Fig. 3. The cubes in a subset are further assigned to SMs of GPUs. The computations for the cells in cubes are assigned into threads, which are executed on SPs. As the computational cost is the same, efficient parallel processing using multiple nodes can be expected.

Besides, the effective use of the shared memory in a GPU is essential to reduce the number of global memory accesses requiring a long access latency. By storing data with high locality to a ring buffer on the shared memory as shown in Fig. 4, it can not only reduce the number of the off-chip memory accesses but also effectively utilize the limited capacity of the on-chip memory.

4 Performance Evaluation and Discussions

The flow simulations using BCM around 3D test models are performed on the supercomputing systems shown in Table 1. F1 is a large model of 200 million cells, and Sphere is a small model of 5 million cells which are shown in Fig. 5.

Figure 6 shows the sustained performance of BCM achieved for the F1 model. The results show that SX-9 achieves a higher sustained performance than the others. As BCM is a memory-intensive application, the sustained memory bandwidth has a great impact on the performance. In addition to the importance of a high memory bandwidth, the effective use of ADB further improves the sustained bytes/flop ratio, resulting in the high performance.

Even though the peak memory bandwidth of FX-1 outperforms those of Nehalem EP and EX, its sustained performance is lower. This is because the sustained memory bandwidth of Nehalem is higher than that of SPARC64VII. In the STREAM benchmark, FX1 achieves only 10.0 GB/s while Nehalem EP and EX achieve 17.0 and 17.6 GB/s, respectively.

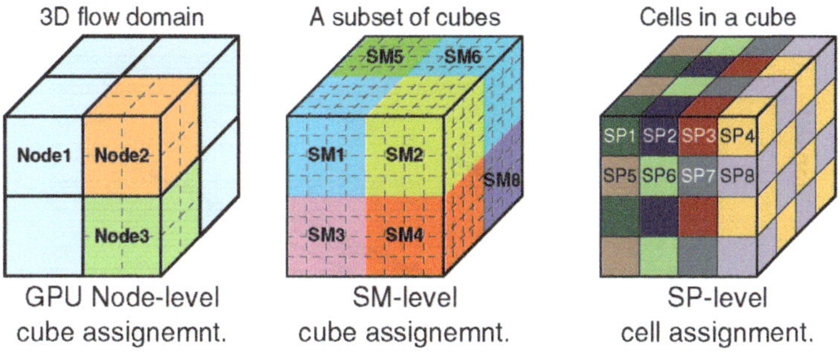

Fig. 3 Task assignments into a GPU cluster system

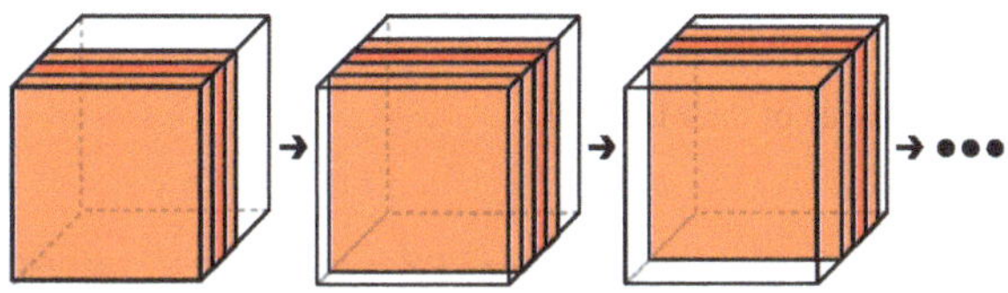

Fig. 4 Three planes for cyclical use of the shared memory

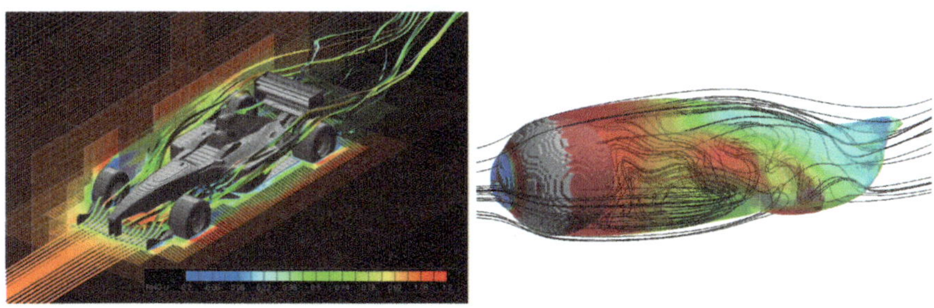

Fig. 5 3D test modes. (*Left*: F1, *Right*: Sphere)

Figure 7 shows the sustained performance of BCM on the Sphere model, which includes the results of the GPU system. This results shows that a GPU cluster system achieves comparable and/or better performance than the scalar cluster systems. The main reason is that GPUs can accelerate the data parallel calculations of BCM using a number of SPs and high memory bandwidth, even though it cannot execute a large problem such as the F1 model due to the limited global memory capacity. Effective use of SPs and shared memory contributes to the good sustained performance larger than the other scalar systems. Another reason is that the sustained performances of other supercomputers including SX-9 become lower for a small problem such as the Sphere model. However, even if the calculations using the GPUs are fast, data transfers between a GPU and a CPU in a node and between GPUs in different nodes

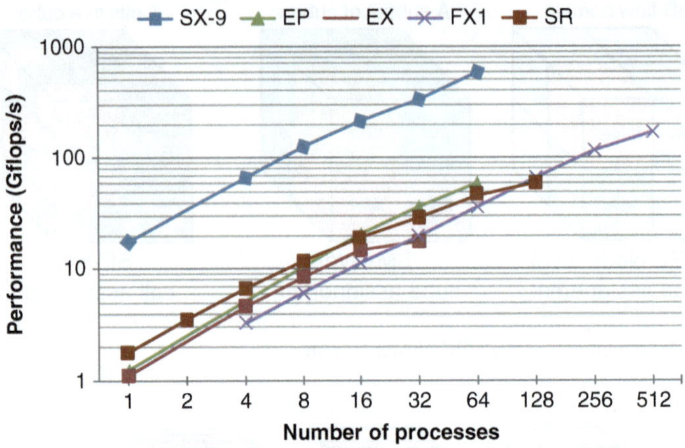

Fig. 6 Sustained performance of the F1 model

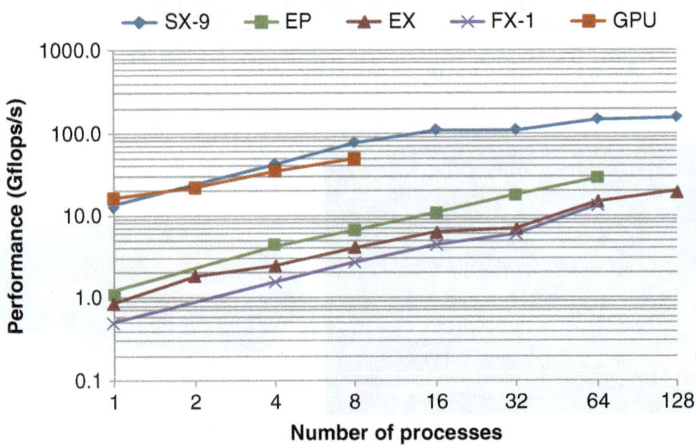

Fig. 7 Sustained performance of the sphere model

are slow and cannot be negligible. As a result, the data transfer dominates the most of time in the simulation. To further accelerate BCM using the GPU system, the time of data transfers should be shortened and be hidden by transferring data during the calculations.

The ratio of the sustained performance to the peak performance on an SX-9 vector processor is about 17 %, while those of the other systems are about 1.5–3.2 %. This is because the vector units in a vector processor are efficiently utilized for the calculations.

Taking a look at the scalability shown in Fig. 8, all of the systems achieve high scalability in the F1 model due to a large number of parallel tasks and sufficient

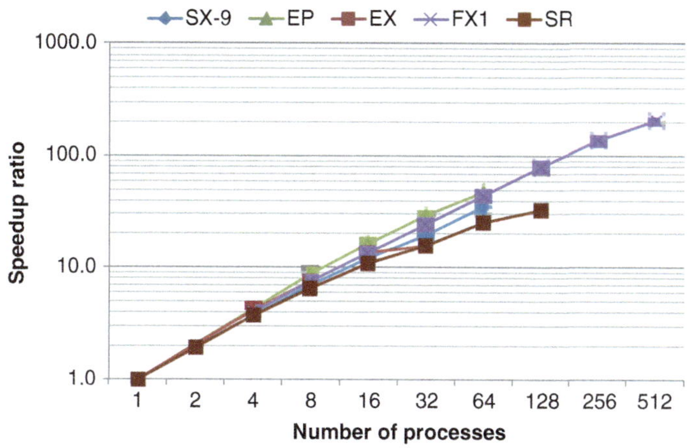

Fig. 8 Speedup ratio of the F1 model

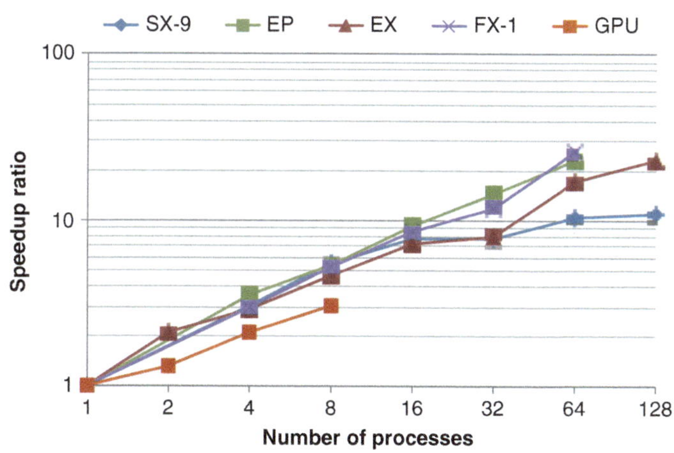

Fig. 9 Speedup ratio of the sphere model

network bandwidth. The scalabilities in the Sphere model shown in Fig. 9 are lower than those of the F1 model. The low scalability of the GPU system comes from the overhead of data transfers. The lower scalabilities of the other systems come from the lack of parallel tasks due to a small number of cubes in the Sphere model.

5 Concluding Remarks

This paper describes the implementations and optimizations of BCM for various types of supercomputing systems such as a scalar type, a vector type, and an accelerator type. The implementation and optimizations considering the characteristics of both a supercomputing system and an application are necessary for high sustained performance. From the performance evaluations of BCM on SX-9, a Nehalem-EP cluster, a Nehalem-EX cluster, FX-1, Hitachi SR16000 M1, and a GPU cluster, it is clarified that the memory bandwidth and the network bandwidth greatly affect the sustained performance of BCM. Therefore, the supercomputing systems that can achieve a high-sustained memory bandwidth and network bandwidth, such as SX-9, are the most promising for further acceleration of BCM.

Acknowledgements The author would like to thank Dr. Kazuhiro Nakahashi of JAXA, Lecturer Daisuke Sasaki of Kanazawa Institute of Technology, Assistant Professor Shun Takahashi of Tokyo University Agriculture and Technology, and Dr. Akihiro Musa of NEC cooperation for valuable discussions on this research. This research was partially supported by Grant-in- Aid for Scientific Research (S) #21226018; Grant-in- Aid for Young Scientists (B) #23700028; Core Research of Evolutional Science and Technology of Japan Science and Technology Agency (JST CREST).

References

1. Dukowicz, J.K.: Approximate factorization as a high order splitting for the implicit incompressible flow equations. Journal of Computational Physics **102**, 336–347 (1992)
2. Ishida, T., Takahashi, S., Nakahashi, K.: Efficient and robust cartesian mesh generation for building-cube method. Journal of Computational Science and Technology **2**(4), 435–445 (2008)
3. Kim, J., Moin, P.: Application of a fractional-step method to incompressible navier-stokes equation. Journal of Computational Physics **59**, 308–323 (1985)
4. Nakahashi, K.: High-density mesh flow computations with pre-/post-data compressions. In: AIAA paper, pp. 2005–4876 (2005)
5. NVIDIA Corporation: NVIDIA CUDA Compute Unified Device Architecture. http://developer.nvidia.com/category/zone/cuda-zone
6. Perot, J.B.: An analysis of the fractional step method. Journal of Computational Physics **108**, 1–58 (1993)
7. Takahashi, S., Ishida, T., Nakahashi, K., Kobayashi, H., Okabe, K., Shimomura, Y., Soga, T., Musa, A.: Study of high resolution incompressible flow simulation based on cartesian mesh. In: AIAA paper: 47th AIAA Aerospace Sciences Meeting, pp. 2009–563 (2009)
8. Takashi, S.: Study of large scale simulation for unsteady flows. Ph.D. thesis, Tohoku University (2009)

Mortar Methods for Single- and Multi-Field Applications in Computational Mechanics

Alexander Popp, Michael W. Gee, and Wolfgang A. Wall

Abstract Mortar finite element methods are of great relevance as a non-conforming discretization technique in various single-field and multi-field applications. In computational contact analysis, the mortar approach allows for a variationally consistent treatment of non-penetration and frictional sliding constraints despite the inevitably non-matching interface meshes. Other single-field and multi-field problems, such as fluid–structure interaction (FSI), also benefit from the increased modeling flexibility provided by mortar methods. This contribution gives a review of the most important aspects of mortar finite element discretization and dual Lagrange multiplier interpolation for the aforementioned applications. The focus is on parallel efficiency, which is addressed by a new dynamic load balancing strategy and tailored parallel search algorithms for computational contact mechanics. For validation purposes, simulation examples from solid dynamics, contact dynamics and FSI will be discussed.

1 Introduction

Mortar finite element methods, and particularly their application to computational contact mechanics and several other single-field and multi-field problems, have seen a great thrust of research over the last decade. In this contribution, we give a review of the most important features of non-conforming finite element discretization

A. Popp (✉) · W.A. Wall
Institute for Computational Mechanics, Technische Universität München, Boltzmannstr. 15, 85747 Garching, Germany
e-mail: popp@lnm.mw.tum.de; wall@lnm.mw.tum.de

M.W. Gee
Mechanics and High Performance Computing Group, Technische Universität München, Boltzmannstr. 15, 85747 Garching, Germany
e-mail: gee@tum.de

M. Resch et al. (eds.), *Sustained Simulation Performance 2012*,
DOI 10.1007/978-3-642-32454-3_12, © Springer-Verlag Berlin Heidelberg 2013

based on mortar methods with a special emphasis on dual Lagrange multiplier interpolation, parallel efficiency of the devised numerical algorithms and high performance computing. Thus, the present work combines and extends different aspects of our previous work in [5, 8–11], to which we refer for technical details, more profound discussion of the presented methods and further numerical examples.

Originally introduced as a domain decomposition technique for spectral elements in [2], mortar methods are nowadays also widely used within finite element formulations for many different problem classes. The first investigations on mortar finite element methods were typically performed for model problems of Laplace operator type, e.g., the Poisson equation, and formulated as non-conforming variational problem with the coupling constraints directly introduced into the global solution space. Yet, an alternative formulation soon became popular, which is not based on a constrained solution space, but rather introduces Lagrange multipliers in the sense of constrained minimization, thus leading to a typical saddle point formulation. Details of both approaches can, for example, be found in [1, 18, 19].

The mortar approach is characterized by an imposition of the occurrent interface constraints in a weak sense and by the possibility to prove its mathematical optimality. This means that suitable inf-sup conditions and *a priori* error estimates for the consistency error and the best approximation error have been established for the most widely used finite element discretizations and for different choices of the discrete Lagrange multiplier space. If considering first-order finite elements as an example, their optimal spatial convergence of order $\mathcal{O}(h^2)$ measured in the L^2-norm is preserved by mortar methods, despite the fact that non-conforming interfaces are involved. Establishing optimal *a priori* error bounds for unilateral contact problems is more intricate due to the typically reduced regularity of the solution. The interested reader is exemplarily referred to [20] and the references therein.

Especially the choice of the discrete Lagrange multiplier space is an essential question with great implications on the actual algorithmic realization of mortar methods. Standard mortar methods suffer from a serious drawback: they generate high computational costs due to the global character of the resulting interface coupling conditions, see [19] for a detailed explanation. However, this issue can be completely resolved by introducing so-called *dual* Lagrange multiplier spaces as proposed in [18], which are constructed based on a biorthogonality condition such that the interface coupling conditions reduce to purely local constraints.

This approach has been applied very successfully to impose interface constraints for finite deformation contact analysis in [10–12, 20], thus allowing for a variationally consistent treatment of non-penetration and frictional sliding conditions despite the inevitably non-matching interface meshes for finite deformations and large sliding motions. Recently, the focus of research in the field of mortar finite element methods and dual Lagrange multiplier interpolation has been extended towards other single-field applications and especially to coupled multiphysics problems. A computational framework for fluid–structure interaction has been proposed in [8] and the combination of FSI and contact interaction for capturing physical phenomena such as elastohydrodynamic lubrication is discussed in [9, 16, 24]. The coupling of several subdomains with non-matching meshes in computational fluid dynamics can also be efficiently carried out with dual mortar methods, see [4].

The remainder of this contribution is organized as follows: Sect. 2 provides an introduction to mortar finite element methods in the exemplary context of mesh tying for nonlinear solid mechanics. The main part in Sect. 3 then describes several aspects of the efficient parallel implementation of mortar methods within a high performance computing framework. Concretely, dynamic load balancing techniques as well as search algorithms for computational contact mechanics will be addressed. With these general findings at hand, different application scenarios ranging from classical mesh tying in solid mechanics to finite deformation contact and coupled problems such as FSI including contact are highlighted in Sect. 4. Finally, Sect. 5 gives a summary and outlook on current and future research directions.

2 Mortar Finite Element Methods

Mesh tying in solid mechanics (also referred to as tied contact) serves as a model problem for the introduction to mortar finite element methods here. The motivation for such mortar mesh tying algorithms is to connect dissimilar meshes in nonlinear solid mechanics in a variationally consistent manner, see also [13] for a comprehensive overview. Reasons for the occurrence of non-matching meshes can be manifold and range from different resolution requirements in the individual subdomains to the use of different types of finite element interpolations.

Without losing generality, only the case of a body with one sole tied contact interface is considered. A generalization to the case of multiple interfaces is however possible without conceptual differences. Figure 1 gives an overview of the general problem setup and introduces some basic notation. The open sets $\Omega_0^{(i)} \subset \mathbb{R}^3$ and $\Omega_t^{(i)} \subset \mathbb{R}^3$, $i = 1, 2$, represent the two subdomains of the contemplated body in the reference and current configuration, respectively. The surfaces $\partial\Omega_0^{(i)}$ are divided into three disjoint subsets $\Gamma_u^{(i)}$, $\Gamma_\sigma^{(i)}$ and $\Gamma_c^{(i)}$, where $\Gamma_u^{(i)}$ is the Dirichlet boundary, $\Gamma_\sigma^{(i)}$ is the Neumann boundary, and $\Gamma_c^{(i)}$ represents the mesh tying interface. Any gaps and overlaps between the subdomains are excluded, i.e., $\Gamma_c^{(1)} \equiv \Gamma_c^{(2)} \equiv \Gamma_c$. The superscript (1) is commonly referred to as slave side of the problem, whereas the superscript (2) denotes the master side.

On each subdomain $\Omega_0^{(i)}$, the initial boundary value problem (IBVP) of finite deformation elastodynamics needs to be satisfied, i.e.,

$$\mathrm{Div}\boldsymbol{P}^{(i)} + \hat{\boldsymbol{b}}_0^{(i)} = \varrho_0^{(i)}\ddot{\boldsymbol{u}}^{(i)} \qquad \text{in } \Omega_0^{(i)} \times [0, T], \tag{1}$$

$$\boldsymbol{u}^{(i)} = \hat{\boldsymbol{u}}^{(i)} \qquad \text{on } \Gamma_u^{(i)} \times [0, T], \tag{2}$$

$$\boldsymbol{P}^{(i)}\boldsymbol{N}^{(i)} = \hat{\boldsymbol{t}}_0^{(i)} \qquad \text{on } \Gamma_\sigma^{(i)} \times [0, T], \tag{3}$$

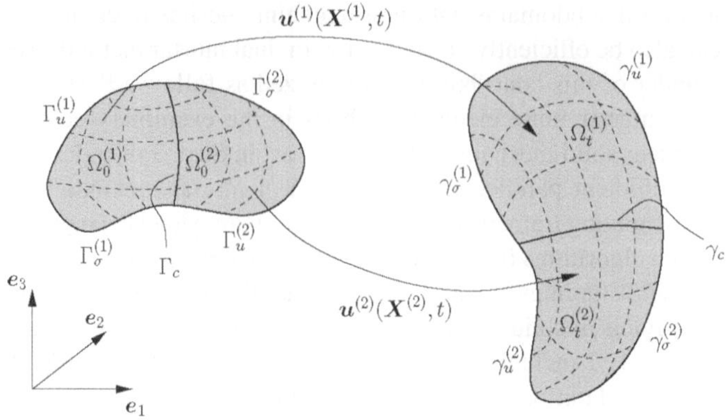

Fig. 1 Configurations, kinematics and basic notation for a mesh tying problem in 3D

$$\boldsymbol{u}^{(i)}(\boldsymbol{X}^{(i)},0) = \hat{\boldsymbol{u}}_0^{(i)}(\boldsymbol{X}^{(i)}) \qquad \text{in } \Omega_0^{(i)}, \tag{4}$$

$$\dot{\boldsymbol{u}}^{(i)}(\boldsymbol{X}^{(i)},0) = \hat{\dot{\boldsymbol{u}}}_0^{(i)}(\boldsymbol{X}^{(i)}) \qquad \text{in } \Omega_0^{(i)}. \tag{5}$$

The mesh tying constraint formulated in the reference configuration is given as

$$\boldsymbol{u}^{(1)} = \boldsymbol{u}^{(2)} \qquad \text{on } \Gamma_c \times [0,T]. \tag{6}$$

Equations (1)–(6) represent the final strong form of a mesh tying problem in nonlinear solid mechanics. In the course of deriving a weak formulations, the balance of linear momentum at the mesh tying interface Γ_c is typically exploited and a Lagrange multiplier vector $\boldsymbol{\lambda}$ is introduced, thus setting the basis for a mixed variational approach. To start the derivation of a weak formulation of (1)–(6), appropriate solution spaces $\mathscr{U}^{(i)}$ and weighting spaces $\mathscr{V}^{(i)}$ need to be defined as

$$\mathscr{U}^{(i)} = \left\{\boldsymbol{u}^{(i)} \in H^1(\Omega) \mid \boldsymbol{u}^{(i)} = \hat{\boldsymbol{u}}^{(i)} \text{ on } \Gamma_u\right\}, \tag{7}$$

$$\mathscr{V}^{(i)} = \left\{\delta\boldsymbol{u}^{(i)} \in H^1(\Omega) \mid \delta\boldsymbol{u}^{(i)} = \boldsymbol{0} \text{ on } \Gamma_u\right\}. \tag{8}$$

Moreover, the Lagrange multiplier vector $\boldsymbol{\lambda} = -\boldsymbol{t}_c^{(1)}$, which represents the *negative* slave side contact traction $\boldsymbol{t}_c^{(1)}$, is chosen from a corresponding solution space denoted as $\mathscr{M}$. In terms of its classification in functional analysis, $\mathscr{M}$ represents the dual space of the trace space $\mathscr{W}^{(1)}$ of $\mathscr{V}^{(1)}$. In the given context, this means that $\mathscr{M} = H^{-\frac{1}{2}}(\Gamma_c)$ and $\mathscr{W}^{(1)} = H^{\frac{1}{2}}(\Gamma_c)$. Based on these considerations, the saddle point type weak formulation is derived next. Basically, this can be done by extending the standard weak formulation of solid mechanics to two subdomains and combining it with Lagrange multiplier coupling terms. Find $\boldsymbol{u}^{(i)} \in \mathscr{U}^{(i)}$ and $\boldsymbol{\lambda} \in \mathscr{M}$ such that

$$\delta \mathscr{W}_{kin}(\ddot{\boldsymbol{u}}^{(i)}, \delta \boldsymbol{u}^{(i)}) + \delta \mathscr{W}_{int,ext}(\boldsymbol{u}^{(i)}, \delta \boldsymbol{u}^{(i)}) + \delta \mathscr{W}_{mt}(\boldsymbol{\lambda}, \delta \boldsymbol{u}^{(i)}) = 0 \quad \forall\, \delta \boldsymbol{u}^{(i)} \in \mathscr{V}^{(i)}, \tag{9}$$

$$\delta \mathscr{W}_{\lambda}(\boldsymbol{u}^{(i)}, \delta \boldsymbol{\lambda}) = 0 \quad \forall\, \delta \boldsymbol{\lambda} \in \mathscr{M}. \tag{10}$$

Herein, the kinetic contribution $\delta \mathscr{W}_{kin}$ as well as the internal and external contributions $\delta \mathscr{W}_{int,ext}$ are well-known from standard weak formulations in nonlinear solid mechanics. The mesh tying interface contribution $\delta \mathscr{W}_{mt}$ and the weak mesh tying constraint $\delta \mathscr{W}_{\lambda}$ have been abbreviated as

$$\delta \mathscr{W}_{mt} = \int_{\Gamma_c} \boldsymbol{\lambda}(\delta \boldsymbol{u}^{(1)} - \delta \boldsymbol{u}^{(2)})\, \mathrm{d}A_0, \qquad \delta \mathscr{W}_{\lambda} = \int_{\Gamma_c} \delta \boldsymbol{\lambda}(\boldsymbol{u}^{(1)} - \boldsymbol{u}^{(2)})\, \mathrm{d}A_0. \tag{11}$$

For the spatial discretization of the tied contact problem (9)–(10), standard isoparametric finite elements are employed. This defines the usual finite dimensional subspaces $\mathscr{U}_h^{(i)}$ and $\mathscr{V}_h^{(i)}$ being approximations of $\mathscr{U}^{(i)}$ and $\mathscr{V}^{(i)}$, respectively. Within our implementation, both first-order and second-order interpolation in 2D and 3D are considered. The subscript h refers to a spatially discretized quantity. Obviously, there exists a direct connection between the employed finite elements in the domains and the resulting surface facets on the mortar interfaces $\Gamma_{c,h}^{(i)}$. For example, a 3D finite element mesh composed of *tet4* and *hex8* elements yields *tri3* and *quad4* facets on the surface of tied contact. Consequently, the following general form of displacement interpolation on the discrete mesh tying surfaces holds:

$$\boldsymbol{u}_h^{(1)}|_{\Gamma_{c,h}^{(1)}} = \sum_{k=1}^{n^{(1)}} N_k^{(1)}(\xi^{(1)}, \eta^{(1)})\, \boldsymbol{d}_k^{(1)}, \quad \boldsymbol{u}_h^{(2)}|_{\Gamma_{c,h}^{(2)}} = \sum_{l=1}^{n^{(2)}} N_l^{(2)}(\xi^{(2)}, \eta^{(2)})\, \boldsymbol{d}_l^{(2)}. \tag{12}$$

The total number of slave nodes on $\Gamma_{c,h}^{(1)}$ is $n^{(1)}$, and the total number of master nodes on $\Gamma_{c,h}^{(2)}$ is $n^{(2)}$. Discrete nodal displacements are given by $\boldsymbol{d}_k^{(1)}$ and $\boldsymbol{d}_l^{(2)}$. The shape functions $N_k^{(1)}$ and $N_l^{(2)}$ are defined with respect to the usual finite element parameter space, usually denoted as $\xi^{(i)}$ for two-dimensional problems (i.e., 1D mesh tying interfaces) and as $\boldsymbol{\xi}^{(i)} = (\xi^{(i)}, \eta^{(i)})$ for three-dimensional problems (i.e., 2D mesh tying interfaces). In addition, an adequate discretization of the Lagrange multiplier vector $\boldsymbol{\lambda}$ is needed, too, and will be based on a discrete Lagrange multiplier space $\mathscr{M}_h$ being an approximation of $\mathscr{M}$. A general notation reads:

$$\boldsymbol{\lambda}_h = \sum_{j=1}^{m^{(1)}} \Phi_j(\xi^{(1)}, \eta^{(1)})\, \boldsymbol{\lambda}_j, \tag{13}$$

with the (still to be defined) shape functions Φ_j and the discrete nodal Lagrange multipliers $\boldsymbol{\lambda}_j$. The total number of slave nodes carrying additional Lagrange multiplier degrees of freedom is $m^{(1)}$. Typically for mortar methods, every slave node also serves as coupling node, and thus in the majority of cases $m^{(1)} = n^{(1)}$ will hold. However, in the context of second-order finite element interpolation it may be favorable to chose $m^{(1)} < n^{(1)}$ in certain cases, see e.g., [12, 20].

Two different families of discrete Lagrange multipliers, namely *standard* and so-called *dual* Lagrange multipliers are commonly distinguished. Standard Lagrange multipliers represent the classical approach for mortar methods (cf. [1, 14]) and lead to identical shape functions for Lagrange multiplier and slave displacement interpolation, i.e., $\Phi_j = N_j^{(1)}$. In contrast, the dual approach is motivated by the observation that the Lagrange multipliers physically represent fluxes (tractions) on the mesh tying interface in the continuous setting. This duality argument is then reflected by constructing dual Lagrange multiplier shape functions based on a so-called biorthogonality condition with the displacements in $\mathscr{W}_h^{(1)}$, see e.g., [18]. While they are in general not continuous and cannot be interpreted as a trace of conforming finite elements, the biorthogonality condition assures that the Lagrange multiplier shape functions Φ_j are again well-defined and satisfy all required approximation properties. One crucial advantage of the dual approach lies in the fact that it heavily facilitates the treatment of typical mortar coupling conditions at the interface, while at the same time preserving the optimality of the method.

Substituting (12) and (13) into the interface virtual work $\delta \mathscr{W}_{mt}$ in (9) yields

$$\delta \mathscr{W}_{mt,h} = \sum_{j=1}^{m^{(1)}} \sum_{k=1}^{n^{(1)}} \boldsymbol{\lambda}_j^{\mathsf{T}} \left(\int_{\Gamma_{c,h}^{(1)}} \Phi_j \, N_k^{(1)} \, \mathrm{d}A_0 \right) \delta \boldsymbol{d}_k^{(1)}$$

$$- \sum_{j=1}^{m^{(1)}} \sum_{l=1}^{n^{(2)}} \boldsymbol{\lambda}_j^{\mathsf{T}} \left(\int_{\Gamma_{c,h}^{(1)}} \Phi_j \, (N_l^{(2)} \circ \chi_h) \, \mathrm{d}A_0 \right) \delta \boldsymbol{d}_l^{(2)}, \tag{14}$$

where $\chi_h : \Gamma_{c,h}^{(1)} \to \Gamma_{c,h}^{(2)}$ defines a suitable discrete mapping from slave to master side of the mesh tying interface. Such a mapping (or projection) becomes necessary due to the fact that the discretized coupling surfaces $\Gamma_{c,h}^{(1)}$ and $\Gamma_{c,h}^{(2)}$ are, in general, no longer geometrically coincident. In (14), nodal blocks of the two mortar integral matrices commonly denoted as $\boldsymbol{D}$ and $\boldsymbol{M}$ can be identified. This leads to the following definitions:

$$\boldsymbol{D}[j,k] = \int_{\Gamma_{c,h}^{(1)}} \Phi_j N_k^{(1)} \mathrm{d}A_0 \, \boldsymbol{I}_{ndim}, \quad j = 1,\ldots,m^{(1)}, \quad k = 1,\ldots,n^{(1)}, \tag{15}$$

$$\boldsymbol{M}[j,l] = \int_{\Gamma_{c,h}^{(1)}} \Phi_j (N_l^{(2)} \circ \chi_h) \, \mathrm{d}A_0 \, \boldsymbol{I}_{ndim}, \quad j = 1,\ldots,m^{(1)}, \quad l = 1,\ldots,n^{(2)}. \tag{16}$$

Note that $\boldsymbol{I}_{ndim} \in \mathbb{R}^{ndim \times ndim}$ is an identity matrix whose size is determined by the global problem dimension $ndim$, i.e., either $ndim = 2$ or $ndim = 3$. In general, both mortar matrices $\boldsymbol{D}$ and $\boldsymbol{M}$ have a rectangular shape, however $\boldsymbol{D}$ becomes a square matrix for the common choice $m^{(1)} = n^{(1)}$. Dual Lagrange

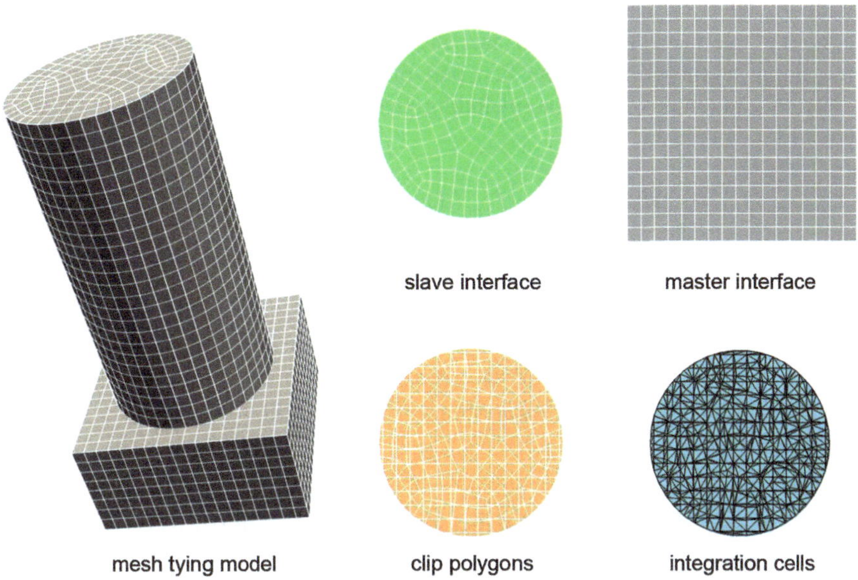

Fig. 2 Main steps of 3D mortar coupling for a mesh tying example—projection and intersection of slave and master surfaces, determination of clip polygons and triangulation into so-called integration cells, see [11, 13] for details

multiplier interpolation based on a biorthogonality condition allows for the beneficial simplification of $\boldsymbol{D}$ reducing to diagonal form. All details concerning the actual numerical integration of the mass matrix type of entries in $\boldsymbol{D}$ and $\boldsymbol{M}$ as well as the implementation of the interface mapping χ_h for both 2D and 3D can be found in our recent work [10, 11]. Here, Fig. 2 only illustrates the generation of integration cells for 3D mortar coupling with an exemplary mesh tying model.

For the ease of notation, all nodes of the two subdomains $\Omega_0^{(1)}$ and $\Omega_0^{(2)}$, and correspondingly all degrees of freedom (DOFs) in the global discrete displacement vector $\boldsymbol{d}$, are sorted into three groups: a group $\mathscr{S}$ containing all slave interface quantities, a group $\mathscr{M}$ of all master quantities and a group denoted as $\mathscr{N}$, which comprises all remaining nodes or DOFs. The global discrete displacement vector can be sorted accordingly, yielding $\boldsymbol{d} = (\boldsymbol{d}_{\mathscr{N}}, \boldsymbol{d}_{\mathscr{M}}, \boldsymbol{d}_{\mathscr{S}})$. Going back to (14), this allows for the following definition:

$$\delta\mathscr{W}_{mt,h} = \delta\boldsymbol{d}_{\mathscr{S}}^{\mathsf{T}}\boldsymbol{D}^{\mathsf{T}}\boldsymbol{\lambda} - \delta\boldsymbol{d}_{\mathscr{M}}^{\mathsf{T}}\boldsymbol{M}^{\mathsf{T}}\boldsymbol{\lambda} = \delta\boldsymbol{d}^{\mathsf{T}}\boldsymbol{B}_{mt}^{\mathsf{T}}\boldsymbol{\lambda} = \delta\boldsymbol{d}^{\mathsf{T}}\boldsymbol{f}_{mt}(\boldsymbol{\lambda}). \tag{17}$$

Herein, the discrete mortar mesh tying operator $\boldsymbol{B}_{mt}$ and the resulting discrete vector of mesh tying forces $\boldsymbol{f}_{mt}(\boldsymbol{\lambda}) = \boldsymbol{B}_{mt}^{\mathsf{T}}\boldsymbol{\lambda}$ acting on slave and master side of the interface are introduced. Due to the saddle point characteristics and resulting symmetry of the mixed variational formulation (9)–(10), the final formulation of the weak constraint contribution $\delta\mathscr{W}_{\lambda}$ can directly be given as

$$\delta\mathscr{W}_{\lambda,h} = \delta\boldsymbol{\lambda}^{\mathsf{T}}\boldsymbol{D}\boldsymbol{d}_{\mathscr{S}} - \delta\boldsymbol{\lambda}^{\mathsf{T}}\boldsymbol{M}\boldsymbol{d}_{\mathscr{M}} = \delta\boldsymbol{\lambda}^{\mathsf{T}}\boldsymbol{B}_{mt}\boldsymbol{d} = \delta\boldsymbol{\lambda}^{\mathsf{T}}\boldsymbol{g}_{mt}(\boldsymbol{d}), \tag{18}$$

with $\boldsymbol{g}_{mt}(\boldsymbol{d}) = \boldsymbol{B}_{mt}\boldsymbol{d}$ representing the discrete mesh tying constraint at the coupling interface. Taking into account the finite element discretization of all remaining contributions to the first part of the weak formulation (9), the semi-discrete equations of motion including tied contact forces and the constraint equations emerge as

$$\boldsymbol{M}\ddot{\boldsymbol{d}} + \boldsymbol{C}\dot{\boldsymbol{d}} + \boldsymbol{f}_{int}(\boldsymbol{d}) + \boldsymbol{f}_{mt}(\boldsymbol{\lambda}) - \boldsymbol{f}_{ext} = \boldsymbol{0}, \tag{19}$$

$$\boldsymbol{g}_{mt}(\boldsymbol{d}) = \boldsymbol{0}. \tag{20}$$

Mass matrix $\boldsymbol{M}$, damping matrix $\boldsymbol{C}$, internal forces $\boldsymbol{f}_{int}(\boldsymbol{d})$ and external forces $\boldsymbol{f}_{ext}$ result from standard FE discretization. It is important to point out, that the actual mortar based interface coupling described here is completely independent of the concrete choice of an underlying finite element formulation.

3 Aspects of Implementation and High Performance Computing

The following section aims at addressing and outlining some of the most important implementation and software design issues associated with the proposed mortar finite element methods. We first put an emphasis on inter-processor redistribution and dynamic load balancing strategies for mortar methods, which in turn requires a short introduction to the employed paradigm of parallel programming. Furthermore, the basic concepts of efficient parallel search algorithms for two body contact, self contact and multiple bodies (e.g., agglomerations of elastic particles) will be presented. All explanations exclusively refer to implementations devised in the context of the present contribution, and subsequently integrated into the in-house finite element software package BACI (cf. [17]), developed at the Institute for Computational Mechanics at Technische Universität München.

3.1 Parallel Redistribution and Dynamic Load Balancing

The presented mortar based mesh tying and contact algorithms are designed for the use on large interconnected computer systems (clusters) with many CPUs and a distributed main memory. Being able to efficiently run large simulations in parallel requires strategies for the partitioning and parallel distribution of the problem data, i.e., finite element meshes (consisting of nodes and elements) as well as global vectors and matrices, into several independent processes. Within BACI, this so-called domain (or data) decomposition is provided by the third-party library ParMETIS, see e.g., [7], and all communication tasks are implemented through MPI.

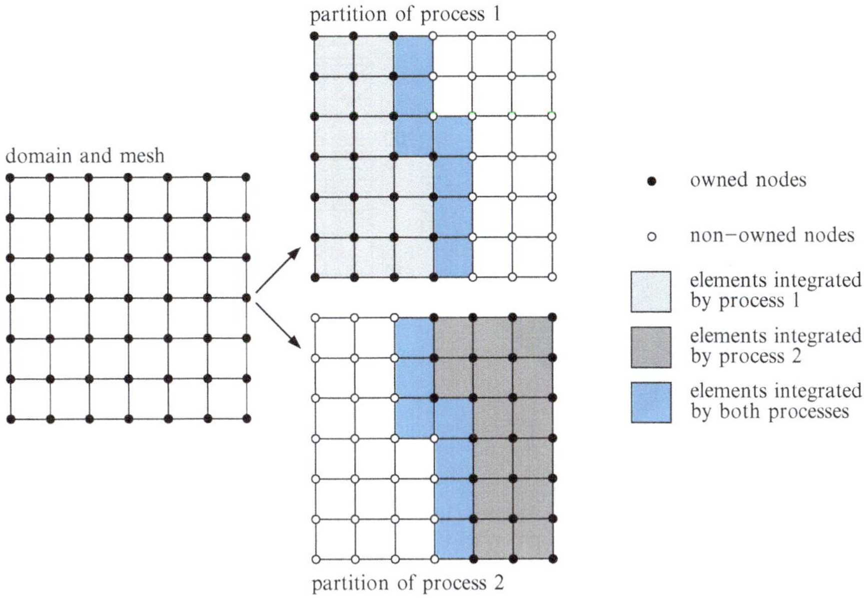

Fig. 3 An example of overlapping domain decomposition and parallel assembly involving two independent processes

An example of such decompositions is visualized in Fig. 3 for a simple partitioning including only two processes. It can be seen that each node in the mesh is uniquely assigned to one specific process, and the same holds true for the elements. In addition, some nodes and elements at the transition between different processes must be stored redundantly within all adjacent processes. Therefore, this type of partitioning is commonly denoted as *overlapping* decomposition. Here, it is sufficient to consider only the most straightforward case of minimal overlap between the individual partitions, i.e., an overlap of one layer of elements or nodes, respectively. Obviously, this concept of overlapping decomposition fits quite naturally to the typical tasks within a finite element program: first, each process performs an elementwise integration of its own partition of the computational domain including the (relatively few) elements at the inter-process boundaries. Then, the resulting quantities (e.g., local element load vectors and stiffness matrices) are assembled into the respective FE nodes of each process. Thus, overlapping domain decomposition as described above provides a very elegant way of processing finite element integration and assembly, which is completely free of communication due to the distributed storage of the resulting global vector and matrix objects. For further details on the C++ based implementation of parallel (i.e., distributed) matrix and vector objects as well as the associated linear algebra, the interested reader is exemplarily referred to the extensive documentation of open-source libraries of the Trilinos Project conducted by Sandia National Laboratories (cf. [6]).

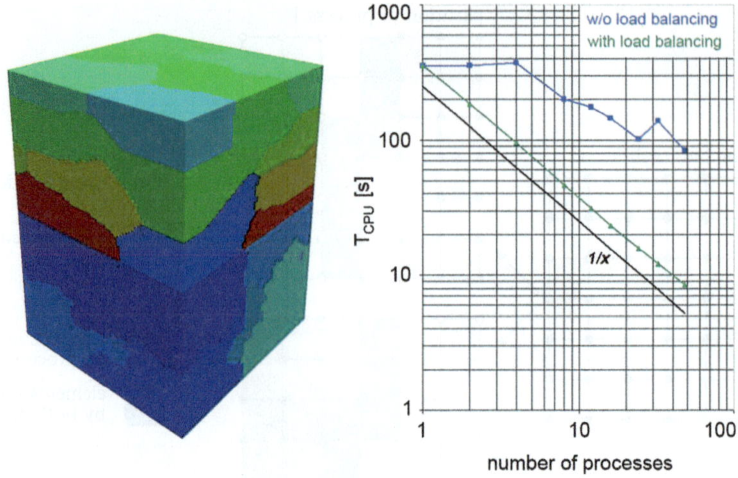

Fig. 4 Parallel redistribution and dynamic load balancing—initial partitioning for exemplary mesh tying problem setup using 32 processes (*left*) and strong scaling diagram (*right*)

Returning to the efficient parallel treatment of mortar methods and the derived mesh tying and contact algorithms, we now examine an exemplary mesh tying problem setup consisting of two cubic bodies, which are distributed in parallel among several processes as depicted in Fig. 4. This partitioning, generated via the ParMETIS library, is in a sense optimal for the integration and assembly of the individual volume finite elements in the two bodies, i.e., the corresponding workload is equally distributed among all processes. For mortar coupling, however, additional (but conceptually similar) tasks have to be performed locally at the mesh tying interface. Computing the mortar contributions to the overall discrete problem formulation involves numerical integration and assembly of the mortar matrices $\boldsymbol{D}$ and $\boldsymbol{M}$, to name only the most important task. Unfortunately, due to its locality, the parallel distribution of the mortar interface itself is not optimal at all.

Figure 4 also illustrates typical results for the parallel efficiency of the presented mortar algorithms in a so-called *strong scaling* diagram. Therein, the CPU time for numerical integration and assembly of all interface-related quantities T_{CPU} is plotted against the total number of processes N with logarithmic scales applied to both axes. Perfect scalability of the examined numerical algorithm is represented by a straight line with a negative slope of -1, thus representing the evident relation that $T_{CPU} \sim 1/N$. It can clearly be seen from Fig. 4 that initially no perfect scalability is achieved with the presented algorithms. This is due to the non-optimal distribution of the slave surface among the participating processes as already described above. The results clearly motivate the need to develop an efficient parallel redistribution and load balancing strategy for mortar finite element methods. The approach proposed in the following is based on three steps, where the first step is of fundamental importance and is therefore needed for both mesh tying and contact applications. In contrast, the second and third step a purely contact-specific.

The rather simple basic idea of the first step is an *independent* parallel distribution of the finite elements in the domain and the mortar elements at the coupling interface in order to achieve optimal parallel scalability of the computational tasks associated with both, i.e., FE integration and assembly in $\Omega^{(1)}$ and $\Omega^{(2)}$ as well as mortar integration and assembly on $\Gamma_c^{(1)}$ and $\Gamma_c^{(2)}$. Again using ParMETIS, this redistribution of the interface elements can readily be performed during problem initialization at $t = 0$. Results for the test model introduced above are also visualized in Fig. 4, thus demonstrating that this simple modification already allows for perfect parallel scalability within a wide range concerning the number of processes. However, dependent on the considered problem size, parallel redistribution only makes sense up to a certain number of processes. It is quite natural that such a limit exists, because there are of course some computational costs associated with the proposed redistribution procedure itself. If too many processes are used in relation to the problem size, these costs (mainly due to communication) become dominant and parallel redistribution is no longer profitable beyond this point.

As already mentioned, this strategy can be further refined for contact applications. In contrast to mesh tying, contact interfaces are characterized by two additional complexities: the actual contact zone is not known *a priori* and it may constantly and significantly vary over time. Thus, in a second and third step, the proposed redistribution strategy is adapted such that it accommodates these additional complexities. Concretely, it can be seen from Fig. 5 that parallel redistribution must be limited to the actual contact area instead of the potential contact area, because the entire computational effort of numerical integration and assembly is connected with the former. Moreover, whenever finite deformations and large sliding motions occur, the redistribution needs to be performed dynamically, i.e., over and over again. Such a dynamic load balancing strategy is then typically triggered by a suitable measure for the workload of each individual process. The parallel balance of the workload among all processes is monitored and a simple criterion whether to apply dynamic load balancing within the current time step or not can be formulated as

$$IF\left(\frac{T_{CPU}^{max}}{T_{CPU}^{min}} > \varepsilon\right) \rightsquigarrow \text{redistribute}. \tag{21}$$

Herein, the minimum and maximum CPU times of one individual process in the last time step are denoted as T_{CPU}^{min} and T_{CPU}^{max}, respectively. The parameter $\varepsilon > 1$ represents a user-defined tolerance. For example, choosing $\varepsilon = 1.2$ implies that at most 20 % unbalance of the parallel workload distribution are tolerated. Of course, the rather simple condition in (21) can easily be extended to incorporate more sophisticated criteria for dynamic load balancing.

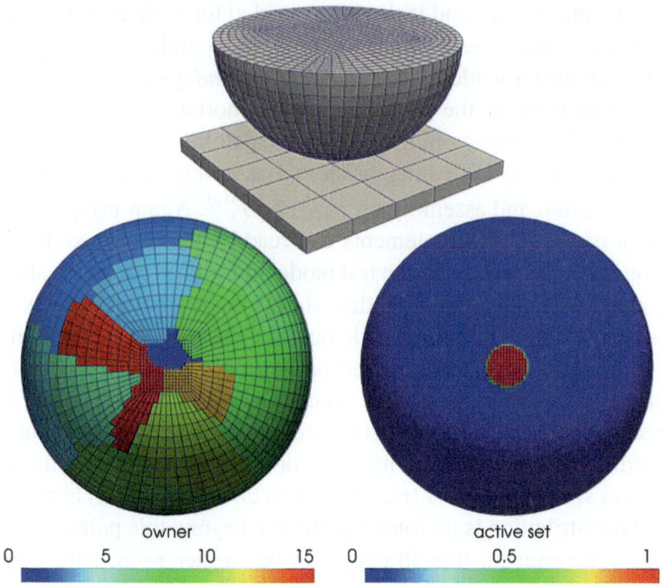

Fig. 5 Motivation for parallel redistribution exemplified with a Hertzian contact example—the active contact region (*bottom right*) is relatively small as compared with the potential contact surface, i.e., the whole hemisphere. Without redistribution only six out of 16 processes would carry the entire workload associated with contact evaluation (*bottom left*)

3.2 Search Algorithms for Two-Body Contact and Self Contact

The search for bodies or individual finite elements that might possible come into contact is an important algorithmic aspect of any FEM contact formulation. In particular, this is true in the context of finite deformations and large sliding motions as primarily considered throughout the present work, because the contact situation continuously changes in such scenarios. Search algorithms have been a subject of intensive research since the beginnings of the computational treatment of contact mechanics problems, see e.g., [21] for a comprehensive overview on the topic.

The basic motivation for efficient contact search algorithms can be easily understood. A naive search approach for two-body contact would require to check *all* finite elements on the slave side against *all* finite elements on the master side for proximity. Thus, the associated number of operations would be $N \times M$, where N and M are the total numbers of slave and master elements, respectively. Assuming $M \approx N$ for the sake of simplicity, the resulting computational complexity of such so-called brute force search algorithms is $\mathcal{O}(N^2)$. Clearly, this makes contact search inacceptably slow already for rather moderate problem sizes.

In the following, a short overview of the parallel search algorithm for two-body contact used in this contribution will be given, which is closely related to the work in [22]. Basically, most contact search algorithms consist of two components, i.e., a

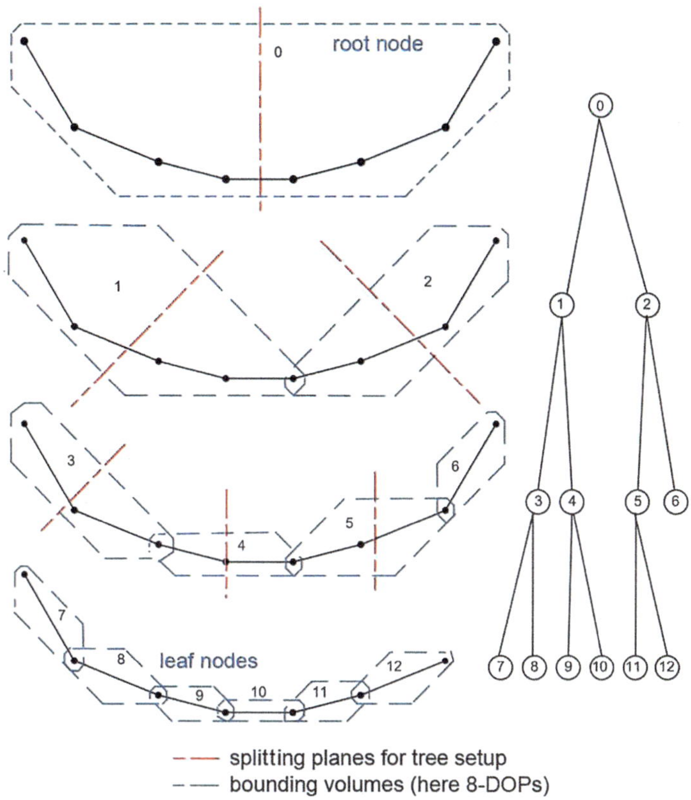

Fig. 6 Search algorithm for two-body contact—two-dimensional example based on 8-DOPs as bounding volumes and a hierarchical binary tree structure

hierarchical global search structure (so-called search tree) and an efficient local geometry representation (so-called bounding volumes). Here, discretized orientation polytopes with k edges (k-DOPs) serve as bounding volumes. Compared to the commonly employed axis-aligned bounding boxes (AABBs), the k-DOPs allow for a much tighter and thus more efficient geometrical representation of the contact surfaces. For 2D simulations, the bounding volumes are typically 8-DOPs, while 18-DOPs are employed in the 3D case. Figure 6 provides a schematic illustration of these ideas for a two-dimensional setting. Further details and comprehensive illustrations can be found in [22]. As can also be seen from Fig. 6, both slave and master surface are then organized and stored within hierarchical binary tree structures, which allow for very fast search and update procedures. The search tree is typically only built once during problem initialization in a top-down way. This process starts from a so-called root node, which contains the entire slave or master contact surface, and then the considered surfaces are continuously divided in halves until arriving at the individual finite elements (so-called leaf nodes of the

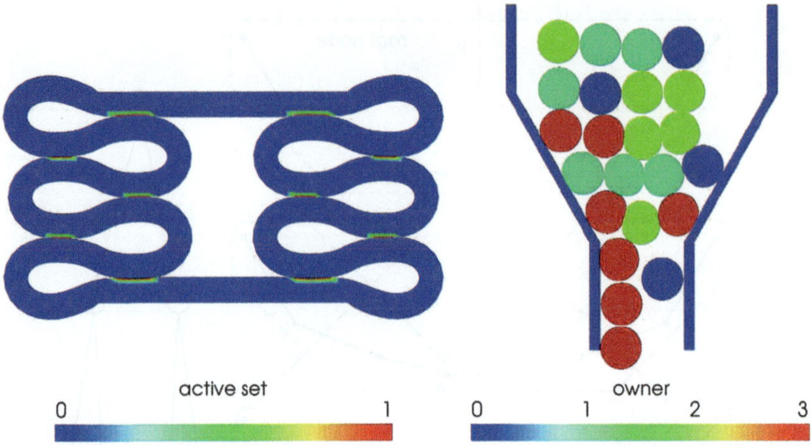

Fig. 7 Examples for self contact and contact of multiple bodies. The active set is visualized for self contact (*left*) and the parallel distribution is shown for the multiple body case (*right*)

search tree). An update of the tree, i.e., of the contact geometry, must be done after each nonlinear iteration step due to the fact finite deformations are considered here.

The search procedure itself basically consists of a recursive algorithm starting with an intersection test of slave and master root nodes. Wherever necessary, i.e., wherever an overlap of the corresponding bounding volumes is detected, the search algorithm proceeds into the lower tree levels until the leaf level is reached. A theoretical analysis in [22] predicts the resulting algorithm complexity to be $\mathcal{O}(N \cdot \log(N^2))$, which has also been confirmed in various numerical investigations. Going beyond the implementation in [22], which is limited to the single-processor case, the presented search algorithm has been extended to fit into a parallel FE simulation framework. As explained above, slave and master surface are then distributed among several independent processes and the tree update as well as search procedures are only performed on the part of the slave contact surface that is actually part of the problem partition of the respective process. This generates a distributed search algorithm with optimal parallel scalability, where only the geometry of the (likewise distributed) master surface needs to be communicated among all processes in order to detect all possible contact pairs.

For the sake of completeness, it is mentioned that two special problem classes in computational contact mechanics, namely self contact and contact of multiple bodies, can be treated within the same algorithmic framework as described above. Two characteristic examples for the mentioned problem classes are illustrated in Fig. 7. The search algorithm for self contact and multiple bodies employed here is again closely related to the work given in [22, 23]. The only algorithmic difference to the two-body case is the way in which the contact search is realized. As can be seen in Fig. 7, self contact is characterized by only one *single* potential self contact surface, so that no a priori definition of slave and master surfaces is possible, but this assignment rather needs to be done in a dynamic manner. Thus, the search procedure needs to be adapted in order to accommodate possible self contact. Moreover, Fig. 7 indicates that the case of multiple bodies lies somewhere in between two-body contact and self contact with regard to its numerical treatment. Again, it is not possible to find a unique a priori definition of slave and master surfaces. However, once slave and master pairs are (dynamically) assigned, this scenario can basically be interpreted and treated as a great many of simple two-body contacts.

4 Exemplary Single-Field and Multi-Field Applications

We present several numerical examples to illustrate the capabilities of the proposed mortar finite element approach. All simulations are based on a parallel implementation of the mesh tying and contact algorithms described above in our in-house research code BACI [17]. The chosen set of examples demonstrates the versatility of mortar finite element methods as non-conforming discretization approach for a manifold of applications in computational mechanics, ranging from single-field problems in solid and fluid mechanics to challenging multi-field problems.

4.1 Mesh Tying in Solid Mechanics

The first numerical example investigates mortar finite element algorithms for 3D mesh tying in the most general context of transient solid dynamics with finite deformations and nonlinear material behavior. As illustrated in Fig. 8, the model consists of an L-shaped block, whose larger part has the dimensions $1.2 \times 1.2 \times 3.6$, while the smaller part is simply a cube with side length 1.2. Constitutive behavior is modeled according to a compressible Neo–Hookean law (Young's modulus $E = 10.000$, Poisson's ratio $\nu = 0.4$), and the density is set to $\varrho_0 = 100$. The left and right surfaces of the L-shaped block are both subject to a pressure load $p(t) = 2{,}000 \times \sin(2\pi t)$ in the time interval $t \in [0, 0.5]$, and the body then moves freely in the time interval $t \in [0.5, 5]$, which adds up to a total of 500 time steps with the step size $\Delta t = 0.01$. A curved non-matching mortar interface is introduced to make the mortar setting as general as possible, with the outer surface of the cylindrical inclusion being chosen as slave side.

In order to assure exact algorithmic conservation of linear and angular momentum as well as mechanical energy, the energy-momentum method (EMM) initially proposed in [15] is employed as time integration scheme here. Some characteristic stages of deformation are also visualized in Fig. 8 and emphasize the strong nonlinearities involved in this simulation. However, the main focus of interest for the presented example lies in the mechanical conservation properties. As can be seen from Fig. 9, linear and angular momentum as well as mechanical energies are *exactly* conserved when combining the proposed dual mortar finite element discretization

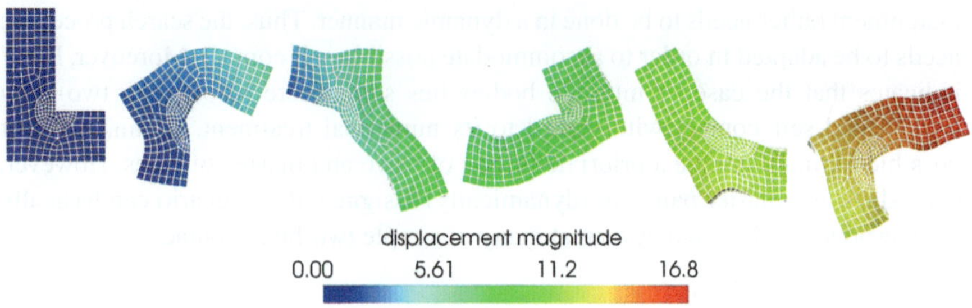

Fig. 8 L-shaped block—problem setup, characteristic stages of deformation at $t = 0, 1, 2, 3, 4, 5$ and numerical solution for the displacement magnitude $\|\boldsymbol{u}\|$

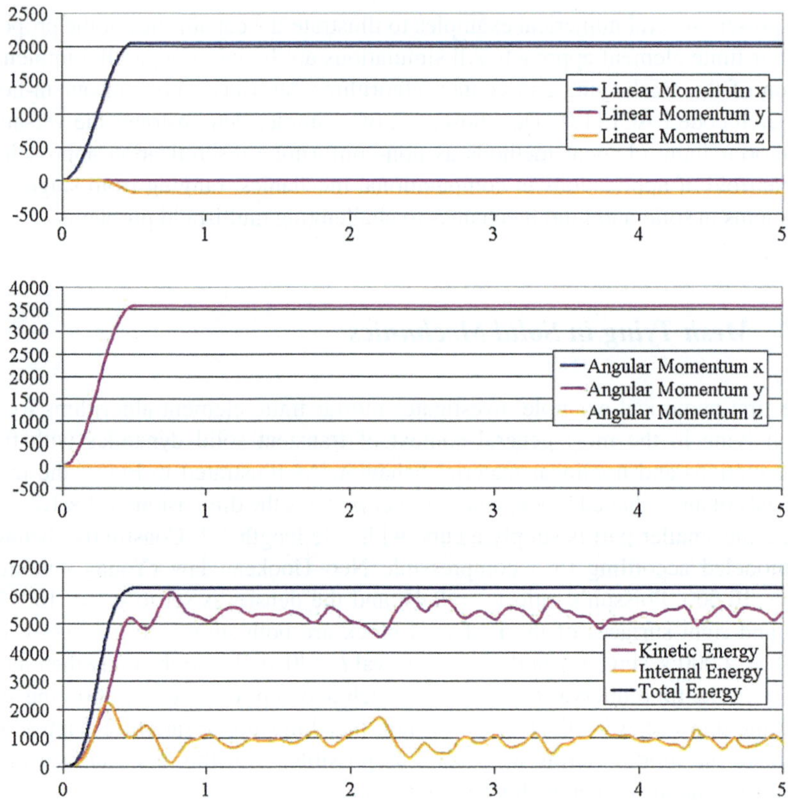

Fig. 9 L-shaped block—exact algorithmic conservation of linear momentum (*top*), angular momentum (*middle*) and mechanical energies (*bottom*) over time t

and EMM time integration. Linear momentum conservation is assured by using the same integration procedure for both mortar matrices $\boldsymbol{D}$ and $\boldsymbol{M}$, while the mesh initialization procedure suggested in [13] and the EMM together guarantee angular momentum conservation. Finally, energy conservation is a direct consequence of

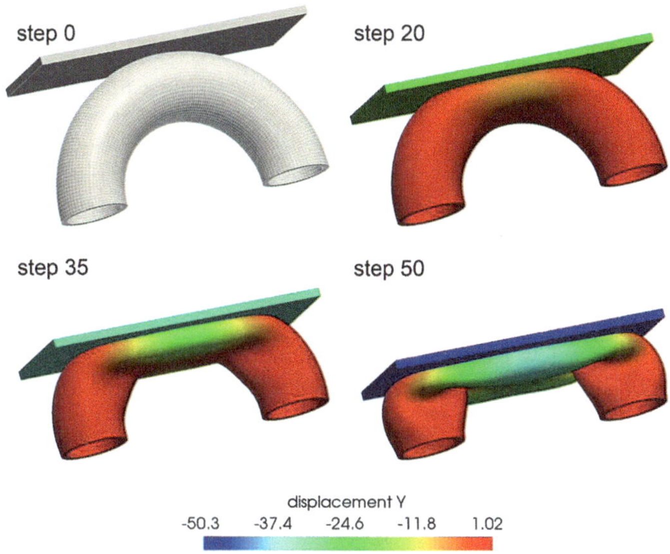

Fig. 10 Torus impact—problem setup and characteristic stages of deformation

the employed time integration scheme, and could not be achieved when using other well-known implicit time integrators such as the generalized-α method [3].

4.2 *Finite Deformation Contact Mechanics*

The second example illustrates the robustness of the described mortar approach in the case of finite deformation contact with significant active set changes. Second-order finite element interpolation in combination with novel, recently proposed dual Lagrange multipliers is employed, see [12, 20] for details.

The considered test setup consists of a hollow half-torus (Neo–Hookean material model with Young's modulus $E = 100$, Poisson's ratio $\nu = 0.3$) and a rigid planar surface. The major and minor radii of the half-torus are 76 and 24, respectively, and the wall thickness is 4.5. The bottom surfaces of the half-torus are completely fixed, and an impact is generated by moving the rigid wall towards the elastic body with a prescribed displacement $u = 50$ accumulated over 50 quasi-static load steps. Figure 10 shows the finite element mesh consisting of 20-node hexahedral elements (with 50.720 nodes in total) as well as some characteristic stages of deformation.

The given impact situation can be considered extremely challenging for any employed active set strategy. However, as Table 1 exemplarily confirms for one representative load step, the semi-smooth Newton type active set strategy presented in [10, 11] and also used here does not have any problems with the described scenario, but resolves all nonlinearities (including the search for the correct active

Table 1 Torus impact—convergence behavior of the semi-smooth Newton method in terms of the relative L^2-norm of the total residual for a representative load step

Step	Relative L^2-norm of residual
1	7.31e+01 (*)
2	6.67e+01 (*)
3	3.54e+01 (*)
4	8.16e+00 (*)
5	4.76e−01
6	8.34e−05
7	9.66e−09

(*) = change in active contact set

set) within only a few Newton–Raphson iteration steps. Owing to the underlying consistent linearization, quadratic convergence is obtained in the limit.

4.3 *Fluid–Structure–Contact Interaction (FSCI)*

The third numerical example presented here demonstrates the effectiveness of the proposed mortar finite element framework for fluid–structure–contact interaction (FSCI) and especially its ability to deal with finite structural deformations and contact in combination with classical fluid–structure interaction (FSI). A variety of problems in engineering and applied sciences require the simulation of unilateral contact of solids surrounded by an incompressible fluid. Important fields of application include machine parts, such as gaskets or sliding-contact bearings, and biomechanical systems, such as heart valves or capillary flow of red blood cells. All details on the computational framework for such coupled problems can be found in [9]. A beam-like structure (Young's modulus $E = 2{,}000$, Poisson's ratio $\nu = 0.4$) is positioned in a two-dimensional channel flow, see Fig. 11. It should be pointed out that the given implementation in BACI is inherently three-dimensional, so that this 2D example is actually modeled as a 3D problem with just one layer of elements in the third direction. A parabolic inflow profile is applied as Dirichlet boundary condition at the left and a zero traction Neumann boundary condition is assumed at the outflow. All remaining channel boundaries are rigid walls with contact occurring between the beam-like structure and a circular obstacle. Exemplarily, 8-node hexahedral elements are used for both the fluid mesh and the structural discretization.

The resulting flow field and structural deformation including contact are illustrated in Fig. 11, giving an impression of this highly dynamic fluid–structure–contact interaction (FSCI) process. The beam-like structure exhibits large deformations: At first, they are primarily induced by fluid stresses resulting from the increasing fluid pressure, i.e., a typical fluid–structure interaction process is initiated. At later stages the gap between beam and obstacle closes completely and the structural deformation is then dominated by contact interaction. It should be mentioned that the given example represents more of a qualitative proof of concept for the successful integration of the mortar contact formulation into a

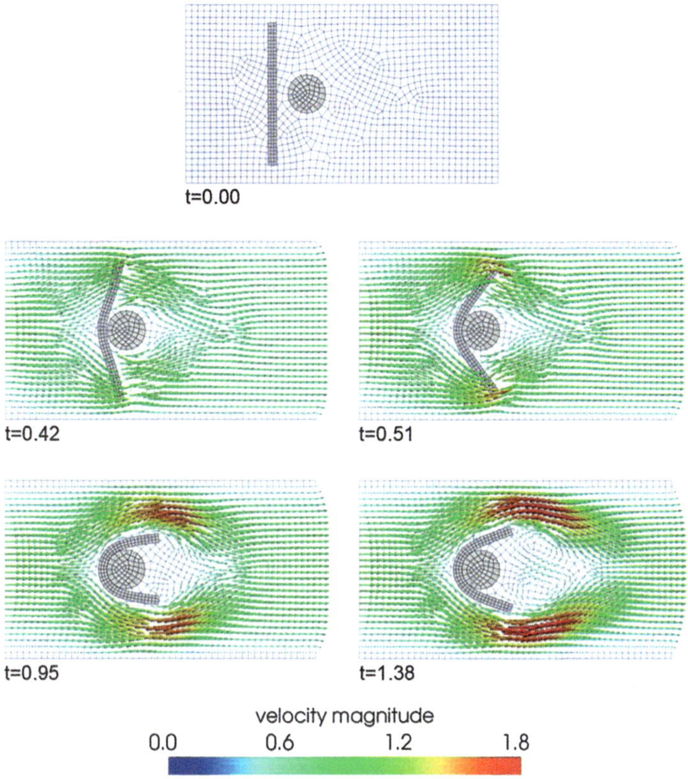

Fig. 11 Beam-like structure in channel flow—finite element mesh (*top*), fluid velocity and structural deformation in several characteristic time steps

fixed-grid FSI framework. Further investigations can also be found in [9]. While the obtained preliminary results are definitely promising towards the simulation of more challenging FSCI applications, the complex physical phenomena occurring during the approach of the two bodies and the associated transition of boundary conditions from FSI type to contact type are not yet fully captured here due to an insufficient fluid mesh resolution.

4.4 *Large-Scale Simulations*

The mortar finite element methods presented in this contribution are readily applicable to large-scale simulations of complex processes involving fluid dynamics, structural dynamics or several coupled physical fields. Especially the design of the numerical algorithms as described in Sect. 3, i.e., including parallel search procedures with hierarchic binary tree structures and dynamic load balancing,

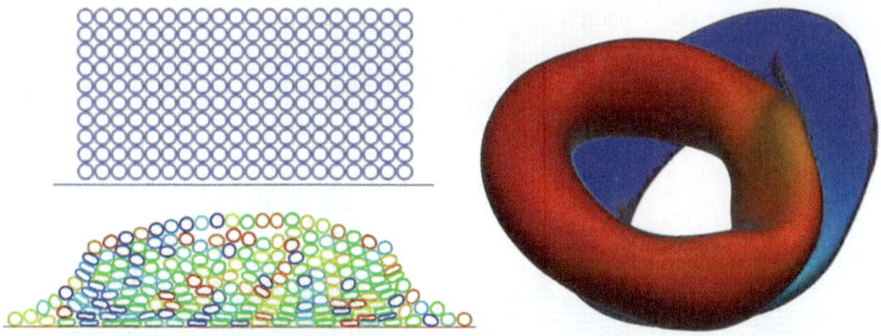

Fig. 12 Exemplary large-scale simulations—contact and self contact of 200 elastic rings in 2D (*left*) and impact of two thin-walled tori in 3D (*right*)

assures an excellent parallel scalability when solving very large simulation models with up to several million degrees of freedom. To give an impression of typical problem sizes, snapshots of two large-scale examples from computational contact dynamics with finite deformations and nonlinear material behavior are shown in Fig. 12. The finite element mesh for the 3D impact model, for example, consists of 4,255,360 hexahedral elements and 13,994,880 degrees of freedom in total. The numerical solution is performed in parallel on up to 120 cores, using an implicit time stepping scheme and 500 time increments to resolve all contact interactions.

5 Conclusions and Outlook

The most important aspects of mortar finite element methods as a non-conforming discretization technique for a wide range of applications in computational mechanics have been reviewed. In particular, several new approaches for the efficient parallel implementation of such mortar methods within a high performance computing framework have been addressed. Contact search algorithms based on hierarchical tree structures and especially a novel dynamic load balancing strategy specifically developed for mortar based discretization have been shown to be indispensable ingredients of efficient numerical algorithms.

Beyond classical applications in nonlinear solid mechanics and contact analysis, the application of mortar finite element methods within a simulation framework for fluid–structure–contact interaction has been addressed shortly. This can be seen as a first step towards more challenging multiphysics and multiscale simulations, which couple contact analysis with several other physical fields and take into account effects on different length scales. Ongoing and future work in this field focuses on modeling finite deformation contact between slender beams and coupled thermomechanical contact, with applications ranging from Brownian dynamics of polymers in fiber networks to heat conduction and mechanical dissipation due to frictional sliding in thermally loaded machine parts.

Acknowledgements The support of the first author (A.P.) by the TUM Graduate School is gratefully acknowledged.

References

1. Ben Belgacem, F.: The mortar finite element method with Lagrange multipliers. Numerische Mathematik **84**(2), 173–197 (1999)
2. Bernardi, C., Maday, Y., Patera, A.T.: A new nonconforming approach to domain decomposition: the mortar element method. In: H. Brezis, J. Lions (eds.) Nonlinear partial differential equations and their applications, pp. 13–51. Pitman/Wiley: London/New York (1994)
3. Chung, J., Hulbert, G.M.: A time integration algorithm for structural dynamics with improved numerical dissipation: The generalized-α method. Journal of Applied Mechanics **60**, 371–375 (1993)
4. Ehrl, A., Popp, A., Gravemeier, V., Wall, W.A.: A mortar approach with dual Lagrange multipliers within a variational mutiscale finite element method for incompressible flow. Computer Methods in Applied Mechanics and Engineering, submitted (2012)
5. Gitterle, M., Popp, A., Gee, M.W., Wall, W.A.: Finite deformation frictional mortar contact using a semi-smooth newton method with consistent linearization. International Journal for Numerical Methods in Engineering **84**(5), 543–571 (2010)
6. Heroux, M.A., Bartlett, R.A., Howle, V.E., Hoekstra, R.J., Hu, J.J., Kolda, T.G., Lehoucq, R.B., Long, K.R., Pawlowski, R.P., Phipps, E.T., Salinger, A.G., Thornquist, H.K., Tuminaro, R.S., Willenbring, J.M., Williams, A., Stanley, K.S.: An overview of the Trilinos project. ACM Transactions on Mathematical Software **31**(3), 397–423 (2005)
7. Karypis, G., Kumar, V.: A parallel algorithm for multilevel graph partitioning and sparse matrix ordering. Journal of Parallel and Distributed Computing **48**(1), 71–95 (1998)
8. Klöppel, T., Popp, A., Küttler, U., Wall, W.A.: Fluid–structure interaction for non-conforming interfaces based on a dual mortar formulation. Computer Methods in Applied Mechanics and Engineering **200**(45–46), 3111–3126 (2011)
9. Mayer, U.M., Popp, A., Gerstenberger, A., Wall, W.A.: 3D fluid–structure–contact interaction based on a combined XFEM FSI and dual mortar contact approach. Computational Mechanics **46**(1), 53–67 (2010)
10. Popp, A., Gee, M.W., Wall, W.A.: A finite deformation mortar contact formulation using a primal-dual active set strategy. International Journal for Numerical Methods in Engineering **79**(11), 1354–1391 (2009)
11. Popp, A., Gitterle, M., Gee, M.W., Wall, W.A.: A dual mortar approach for 3D finite deformation contact with consistent linearization. International Journal for Numerical Methods in Engineering **83**(11), 1428–1465 (2010)
12. Popp, A., Wohlmuth, B.I., Gee, M.W., Wall, W.A.: Dual quadratic mortar finite element methods for 3D finite deformation contact. SIAM Journal on Scientific Computing, accepted (2012)
13. Puso, M.A.: A 3D mortar method for solid mechanics. International Journal for Numerical Methods in Engineering **59**(3), 315–336 (2004)
14. Seshaiyer, P., Suri, M.: hp submeshing via non-conforming finite element methods. Computer Methods in Applied Mechanics and Engineering **189**(3), 1011–1030 (2000)
15. Simo, J.C., Tarnow, N.: The discrete energy-momentum method. Conserving algorithms for nonlinear elastodynamics. Zeitschrift für Angewandte Mathematik und Physik **43**(5), 757–792 (1992)
16. Stupkiewicz, S.: Finite element treatment of soft elastohydrodynamic lubrication problems in the finite deformation regime. Computational Mechanics **44**(5), 605–619 (2009)
17. Wall, W.A., Gee, M.W.: BACI: A multiphysics simulation environment. Tech. rep., Technische Universität München (2012)

18. Wohlmuth, B.I.: A mortar finite element method using dual spaces for the Lagrange multiplier. SIAM Journal on Numerical Analysis **38**(3), 989–1012 (2000)
19. Wohlmuth, B.I.: Discretization methods and iterative solvers based on domain decomposition. Springer-Verlag Berlin Heidelberg (2001)
20. Wohlmuth, B.I., Popp, A., Gee, M.W., Wall, W.A.: An abstract framework for a priori estimates for contact problems in 3D with quadratic finite elements. Computational Mechanics, DOI: 10.1007/s00466-012-0704-z (2012)
21. Wriggers, P.: Computational contact mechanics. John Wiley & Sons (2002)
22. Yang, B., Laursen, T.A.: A contact searching algorithm including bounding volume trees applied to finite sliding mortar formulations. Computational Mechanics **41**(2), 189–205 (2008)
23. Yang, B., Laursen, T.A.: A large deformation mortar formulation of self contact with finite sliding. Computer Methods in Applied Mechanics and Engineering **197**(6–8), 756–772 (2008)
24. Yang, B., Laursen, T.A.: A mortar-finite element approach to lubricated contact problems. Computer Methods in Applied Mechanics and Engineering **198**(47–48), 3656–3669 (2009)

Massive Computation for Femtosecond Dynamics in Condensed Matters

Yoshiyuki Miyamoto

Abstract In this report, numerical simulation on non-thermal dynamics in condensed matters conducted by femtosecond laser shot is presented. Electron–ion dynamics was treated within the framework of the time-dependent density functional theory for electrons coupled with classical molecular dynamics for ions. The formalisms and application of this simulation to photo-exfoliation of graphene from graphite surface and photo-disintegration of molecules inside a carbon nanotube are presented. Heavy tasks for memory access in this simulation scheme will also be mentioned.

1 Introduction

Recent development of material fabrication using femtosecond laser [1] requires advanced theoretical approach that can simulate electron–ion dynamics. This theoretical approach should be beyond the perturbation theory that treats static solutions for both ground and excited states. The computational scheme treating real-time propagation of electron being coupled with classical molecular dynamics (MD) of ion is relevant to simulate laser-induced structural change. Thanks to development of high-performance computers, such simulation will become useful tool to conduct experimental frontier researches. This paper reviews theoretical background of this simulation in Sect. 2, and applications of this simulation scheme are presented in Sect. 3. Then, some notes on numerical tasks needed on high-performance computer are mentioned in Sect. 4. Finally, summary is given in Sect. 5.

Y. Miyamoto (✉)
Nanosystem Res. Institute, National Institute of Advanced Industrial Science and Technology (AIST), Central 2, 1-1-1 Umezono, Tsukuba, 305-8568, Japan
e-mail: yoshi-miyamoto@aist.go.jp

M. Resch et al. (eds.), *Sustained Simulation Performance 2012*,
DOI 10.1007/978-3-642-32454-3_13,

2 Theoretical Backgrounds

In this section, the theoretical background of quantum mechanics based on the density functional theory (DFT) is shown. The DFT is applied to condensed matters with the static treatment of the electronic ground states. Meanwhile, the dynamical extension of DFT enables us to treat non-equilibrium dynamics triggered by electronic excitation with some approximations. A framework on the simulation with presence of intense laser field is also given.

2.1 *Static Treatment*

We can compute static states of electrons in a condensed matter by applying Euler equation to minimize the total energy as a functional of electron wave functions. Within the DFT we can define total energy of condensed matters with a following equation:

$$\begin{aligned} E_{tot} &= \hat{T} + \hat{V}_{nl} + \frac{1}{2}\int\int \frac{\rho(\boldsymbol{r})\rho(\boldsymbol{r}')}{\mid \boldsymbol{r}-\boldsymbol{r}'\mid} d\boldsymbol{r}d\boldsymbol{r}' \\ &+ \int E_{XC}[\rho]\rho(\boldsymbol{r})d\boldsymbol{r} + \sum_I Z_I \int \frac{\rho(\boldsymbol{r})}{\mid \boldsymbol{r}-\boldsymbol{R}_I\mid} d\boldsymbol{r} + V_{ions}. \end{aligned} \tag{1}$$

Here, $\boldsymbol{r}$ means real-space mesh points and $\rho(\boldsymbol{r})$ means valence electron density which is given by,

$$\rho(\boldsymbol{r}) \equiv \sum_n f_n \psi_n^{KS}(\boldsymbol{r})^* \psi_n^{KS}(\boldsymbol{r}), \tag{2}$$

where $\psi_n^{KS}(\boldsymbol{r})$ is called as Kohn–Sham wave function representing valence electron wave function at level n and position $\boldsymbol{r}$ within single-particle representation. Meanwhile f_n is either 0 or 1 representing occupation of the state at level n. The third and fourth terms of Eq. (1) are electron–electron Coulomb repulsion energy and exchange-correlation many-body interaction within DFT. Electron–ion Coulomb attraction energy is the fifth term, with Z_I and $\boldsymbol{R}_I$ representing charge and coordinates of ion I, respectively. The last term is ion–ion Coulomb repulsion term. $\hat{T}$ and $\hat{V}_{nl}$ in Eq. (1) are kinetic energies of all valence electrons and non-local parts of electron–ion interaction. These are respectively written as,

$$\hat{T} \equiv \sum_n f_n \int \psi_n^{KS}(\boldsymbol{r})^*(-\frac{1}{2}\frac{\partial^2}{\partial \boldsymbol{r}^2})\psi_n^{KS}(\boldsymbol{r})d\boldsymbol{r} \tag{3}$$

$$\hat{V}_{nl} \equiv \sum_{\tau,l}\int\int \psi_n^{KS}(\boldsymbol{r})V_{nl}^{\tau,l}(\boldsymbol{r},\boldsymbol{r}')\psi_n^{KS}(\boldsymbol{r}')d\boldsymbol{r}d\boldsymbol{r}'. \tag{4}$$

The $V_{nl}^{\tau,l}(\boldsymbol{r},\boldsymbol{r}')$ imitates interaction between core electrons of ions τ and valence wave functions having a component with angular momentum l around the ion τ. The functional derivative of the total energy (1) with conjugate of the Kohn–Sham wave function $\psi_n^{KS}(\boldsymbol{r})$ gives,

$$\begin{aligned}\frac{\delta E_{tot}}{\delta \psi_n^{ks}(\boldsymbol{r})^*} &= -\frac{1}{2}\frac{\partial^2}{\partial \boldsymbol{r}^2}\psi_n^{KS}(\boldsymbol{r}) + \sum_{\tau,l}\int V_{nl}^{\tau,l}(\boldsymbol{r},\boldsymbol{r}')\psi_n^{KS}(\boldsymbol{r}')d\boldsymbol{r}' \\ &\quad + \int \frac{\rho(\boldsymbol{r}')}{|\boldsymbol{r}-\boldsymbol{r}'|}d\boldsymbol{r}'\psi_n^{KS}(\boldsymbol{r}) + \frac{\delta(E_{XC}\rho)}{\delta\rho}\psi_n^{KS}(\boldsymbol{r}) \\ &\quad + \sum_I \frac{Z_I}{|\boldsymbol{r}-\boldsymbol{R}_I|}\psi_n^{KS}(\boldsymbol{r}) \\ &\equiv H_{KS}(\boldsymbol{r})\psi_n^{KS}(\boldsymbol{r}).\end{aligned} \tag{5}$$

Here we define the Kohn–Sham Hamiltonian $H_{KS}(\boldsymbol{r})$. The total energy minimum condition with restriction of constant total number of all valence electron ($= \sum_n f_n \int \psi_n^{KS}(\boldsymbol{r}')^* \psi_n^{KS}(\boldsymbol{r})d\boldsymbol{r}$) makes the Euler equation as following Kohn–Sham equation

$$H_{KS}(\boldsymbol{r})\psi_n^{KS}(\boldsymbol{r}) = \epsilon_n \psi_n^{KS}(\boldsymbol{r}) \tag{6}$$

with ϵ_n as Lagrange polynomial (the Kohn–Sham eigenvalue) [2]. At the ground state, the self-consistent solution with one-to-one relation between $\rho(\boldsymbol{r})$ and external potential (atomic positions) was proven [3].

2.2 Dynamical Treatment

When we introduce time dependence on charge and the Kohn–Sham wavefunctions respectively as $\rho(\boldsymbol{r},t)$ and $\psi_n^{KS}(\boldsymbol{r},t)$ in Eqs. (1)–(5), the one-to-one relation between time-varying $\rho(\boldsymbol{r},t)$ and time-varying external field (atomic positions) can be proven [4]. This extension is the time-dependent density functional theory (TDDFT). In the time-dependent problem, the action minimum principle instead of the total energy minimization is applied that derives the time-dependent Kohn–Sham equation [4] as

$$i\hbar\frac{d\psi_n^{KS}(\boldsymbol{r},t)}{dt} = H_{KS}(\boldsymbol{r},t)\psi_n^{KS}(\boldsymbol{r},t). \tag{7}$$

Numerical solution of this time-dependent equation must satisfy orthonormal relation of each Kohn–Sham equation as,

$$\int \psi_m^{KS}(\boldsymbol{r},t)^* \psi_n^{KS}(\boldsymbol{r},t) d\boldsymbol{r} = \delta_{m,n}, \tag{8}$$

here δ represents the Kronecker's delta. For performing numerical integration of Eq. (7) along with the time-axis, a unitary operator is useful having a form

$$\psi_n^{KS}(\boldsymbol{r},t+\delta t) = e^{\frac{1}{i\hbar}\int_t^{t+\delta t} H_{KS}(\boldsymbol{r},t')dt'} \psi_n^{KS}(\boldsymbol{r},t) \simeq \prod_i^N e^{\frac{1}{i\hbar} H_{KS}(\boldsymbol{r},t+\delta_i t)} \psi_n^{KS}(\boldsymbol{r},t), \tag{9}$$

in which the integral along time axis in the exponent shown in the middle is approximated as product of operators at discrete values of time from t to $t+\delta t$ shown at the right end. However, the exponential of the Kohn–Sham Hamiltonian $H_{KS}(\boldsymbol{r},t)$ in Eq. (9) cannot be calculated without some approximations. The difficulty is that operators in $H_{KS}(\boldsymbol{r},t)$, see Eq. (5), are not commutable to each other. One of accurate approximation keeping unitary nature of the time-evolution is a split operator technique (Suzuki–Trotter formalisms) [5] in which multiple product of each operator included in $H_{KS}(\boldsymbol{r},t)$ is used. For example, the approximation which omits third order of δt is

$$e^{xH_{KS}(\boldsymbol{r},t)} \simeq e^{\frac{x}{2}\left(-\frac{1}{2}\frac{\partial^2}{\partial r^2}\right)} \left(\prod_{\tau,l} e^{\frac{x}{2}V_{nl}^{\tau,l}}\right) e^{xV_{loc}} \left(\prod_{\tau',l'} e^{\frac{x}{2}V_{nl}^{\tau',l'}}\right) e^{\frac{x}{2}\left(-\frac{1}{2}\frac{\partial^2}{\partial r^2}\right)}, \tag{10}$$

in which $x = \frac{\delta t}{i\hbar}$, and V_{loc} represents all local potential as

$$V_{loc} = \int \frac{\rho(\boldsymbol{r}',t)}{|\,\boldsymbol{r}-\boldsymbol{r}'\,|} d\boldsymbol{r}' + \frac{\delta(\rho E_{XC})}{\delta\rho(\boldsymbol{r},t)} + \sum_I \frac{Z_I}{|\,\boldsymbol{r}-\boldsymbol{R}_I(t)\,|}. \tag{11}$$

The suffices τ, l (τ', l') in the products of operators $e^{\frac{x}{2}V_{nl}^{\tau,l}}$ on the right and left to $e^{xV_{loc}}$ in Eq. (10) should appear in opposite order. This second-order splitting was derived no matter how many operators are included in the $H_{KS}(\boldsymbol{r},t)$. And multiproduct of Eq. (10) with certain weight factor on δt can reach higher-order of accuracy [5] that can be applied for right end of Eq. (9).

Thanks to preserving the unitary nature of the split-operators, we can compute time-evolution of Kohn–Sham wave function $\psi_n^{KS}(\boldsymbol{r},t)$ in parallel with respect to index n without communications of the wave function among used processors. Meantime, the partial charge density $\rho_n(\boldsymbol{r},t) = \psi_n^{KS}(\boldsymbol{r},t)^* \psi_n^{KS}(\boldsymbol{r},t)$ should be reduced to one processor to build total charge $\rho(\boldsymbol{r},t)$, that must be distributed to all processors by using `MPI_All_reduce`. The communication pattern is illustrated in Fig. 1.

Implementation of the Suzuki–Trotter formula for real-time propagation of the Kohn–Sham orbital $\psi_n^{KS}(\boldsymbol{r},t)$ was done being coupled with classical MD [6] using the plane-wave basis set to describe valence electron wave functions. For performing

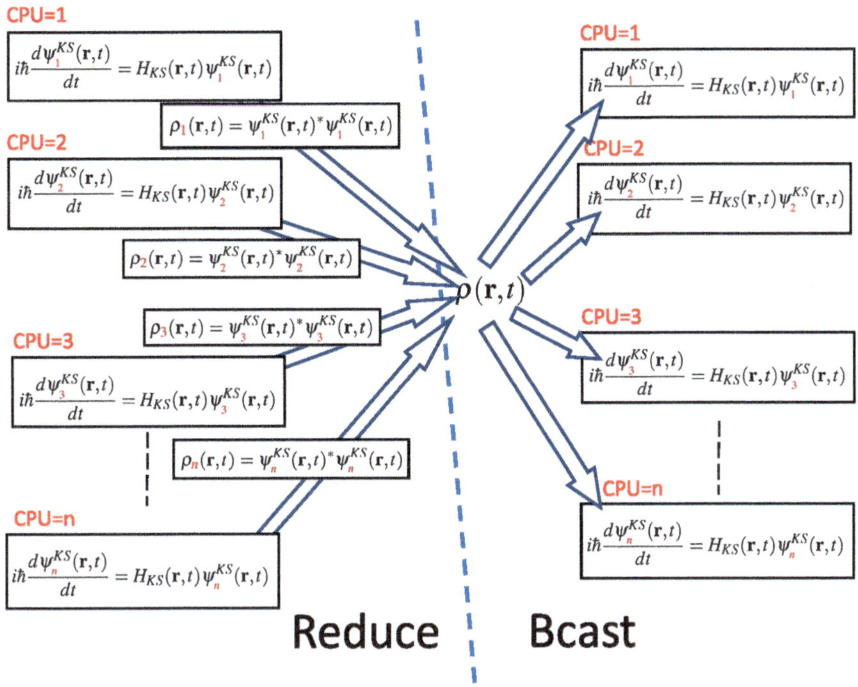

Fig. 1 Parallelizing time-dependent Kohn–Sham equation. (*Left*) Norms of wave functions are reduced to a total charge, which is broadcasted (*right*). These two step are done at every time-step by `MPI_All_reduce`

MD, the force acting on ion I, $\boldsymbol{F}_I$, is approximated as the Hellmann–Feynman force,

$$\boldsymbol{F}_I = -\sum_n f_n \left(\sum_{I,l} \int \int \psi_n^{KS}(\boldsymbol{r}',t)^* \frac{\partial V_{nl}^{I,l}(\boldsymbol{r}',\boldsymbol{r})}{\partial \boldsymbol{R}_I} \psi_n^{KS}(\boldsymbol{r},t) d\boldsymbol{r}' d\boldsymbol{r} \right)$$
$$+ \sum_I Z_I \int \frac{\rho(\boldsymbol{r},t)}{|\,\boldsymbol{r}-\boldsymbol{R}_I\,|^3}(\boldsymbol{r}-\boldsymbol{R}_I)d\boldsymbol{r} + \sum_{I \neq J} \frac{Z_I Z_J}{|\,\boldsymbol{R}_I - \boldsymbol{R}_J\,|^3}(\boldsymbol{R}_I - \boldsymbol{R}_J). \quad (12)$$

Applying the Hellmann–Feynman force in MD is based on Ehrenfest approximation [7], which is equivalent to follow a trajectory of an average of many potential energy surfaces and thus called as the mean-field approximation.

2.3 *Simulation with Intense Laser Field*

Interaction of laser field on electronic system can be described either in Lorentz-gauge or in Coulomb gauge. In Lorentz gauge the momentum operator in the

Hamiltonian $H_{KS}(\boldsymbol{r},t)$ is rescaled with vector potential, while in Coulomb gauge a time-varying scalar potential $V_{ext}(\boldsymbol{r},t)$ is added to the Hamiltonian leading to the equation,

$$i\hbar\frac{d\psi_n^{KS}(\boldsymbol{r},t)}{dt} = [H_{KS}(\boldsymbol{r},t) + V_{ext}(\boldsymbol{r},t)]\,\psi_n^{KS}(\boldsymbol{r},t). \tag{13}$$

Merit of the Coulomb gauge is ability to treat interaction of laser field on both electrons and ions. Since practical application with use of plane-wave basis set requires the three-dimensional periodic boundary conditions, the time-varying scalar potential should also have the same periodicity.

In this work, we introduce time-varying fictitious charge $\rho_{ext}(\boldsymbol{r},t)$ to express the time-varying scalar potential as

$$V_{ext}(\boldsymbol{r},t) = \int \frac{\rho_{ext}(\boldsymbol{r}',t)}{\mid \boldsymbol{r}' - \boldsymbol{r} \mid} d\boldsymbol{r}'. \tag{14}$$

So the total energy was modified from Eq. (1) as

$$E_{tot}(t) = \hat{T} + \hat{V}_{nl} + \frac{1}{2}\int\int \frac{(\rho(\boldsymbol{r}',t) + \rho_{ext}(\boldsymbol{r}',t))\,(\rho(\boldsymbol{r},t) + \rho_{ext}(\boldsymbol{r},t))}{\mid \boldsymbol{r}' - \boldsymbol{r} \mid} d\boldsymbol{r}'d\boldsymbol{r}$$

$$+\int E_{XC}(\rho)\rho(\boldsymbol{r},t)d\boldsymbol{r} + \sum_I Z_I \int \frac{\rho(\boldsymbol{r},t) + \rho_{ext}(\boldsymbol{r},t)}{\mid \boldsymbol{R}_I(t) - \boldsymbol{r} \mid} d\boldsymbol{r} + V_{ions}. \tag{15}$$

One can note that functional derivative of Eq. (15) with respect to conjugate of the Kohn–Sham orbital $\psi_n^{KS}(\boldsymbol{r},t)$ gives $H_{KS}(\boldsymbol{r},t) + V_{ext}(\boldsymbol{r},t)$. Thus Eq. (13) is consistent with the action minimum principle.

In order to investigate the structural change of condensed matter upon irradiation with laser shot, combination of MD simulation and electron dynamics using Eq. (13) is useful. An important thing is numerical stability throughout the simulation no matter what level of approximation is adapted for E_{XC}. When we omit time-to-time correlation in the E_{XC} term of Eq. (15), we can numerically derive work done by laser field $W(t)$ and conservation of

$$E_{tot}(t) + \sum_I \frac{1}{2M_I}\left(\frac{d\boldsymbol{R}_I(t)}{dt}\right)^2 - W(t) \tag{16}$$

can be checked, in which M_I is mass of ion I. This conservation rule was introduced by performing the numerical integration of the time-derivative of $W(t)$, that can be numerically computed along time-axis [8].

3 Applications

In this section, two examples of application of the TDDFT-MD simulation on photo-induced structural change in condensed matters are presented. The first one is exfoliation of graphene layer from the surface of graphite upon irradiation with femtosecond laser shot. The considered wavelength of the laser is 800 nm and the pulse duration has full-width of half-maximum (FWHM) of 45 fs. This computational tests can design a way of fabricating graphene sheet, which was isolated recently [9] and attracted much interests from scientific and industrial views due to its unique electronic and thermal properties. The following simulation would provide ways of graphene formation which can be alternative to mechanical peeling off [9] or chemical vapor depositions on metal [10, 11] or SiC [12] substrates.

Another example of TDDFT-MD simulation is manipulating molecules encapsulated inside carbon nanotubes (CNTs) with very short pulse with FWHM as 2 fs. After discovery of capillary effect of CNTs [16], molecular encapsulation inside CNT was intensively studied, and optical absorption of β-carotene [17] was reported recently. This report raised an interest on feasibility of optical penetration of "black-colored" nanotube to excite molecules inside. The simulation mentioned in following shows feasibility of manipulating molecules inside semiconducting CNT with very short pulse laser having FWHM as 2 fs.

3.1 Laser Exfoliation of Graphene from Graphite

Ultra-fast measurement of electron-beam diffraction technique for time-evolution of inter-layer distance of bulk graphite was done under irradiation with the femtosecond laser [13, 14], and early time contraction and later time expansion of interlayer distance was observed. Yet this diffraction technique can monitor structural change solely of the bulk region of graphite and the surface region remained unknown. Meanwhile, a simulation giving electron temperature [15] needed to assume semi-equilibrium condition which may not hold in ultra-fast dynamics within hundreds femtoseconds. In recent work [18], we have performed TDDFT-MD simulation showing spontaneous exfoliation of surface mono-layer of graphene from graphite. The assumed wavelength of the femtosecond laser is 800 nm, FWHM is 45 fs. According to the experimental condition of laser polarization (p-polarization) [13, 14], we assume laser field is normal to the graphite surface. Under the periodic boundary conditions, spatial variation of the scalar potential $V_{ext}(\boldsymbol{r}, t)$ is linear but rapidly change the polarity at the vacuum region far from the graphite region. The numerical details were shown in the former work [18].

Figure 2 shows pulse shape of currently assumed femtosecond laser shot and corresponding result of TDDFT-MD simulation upon irradiation with this pulse. One can note that the structural change is not noticeable even at the vanishing of laser field (at 80 fs), while much later (at 161 fs), only topmost layer leaves [18].

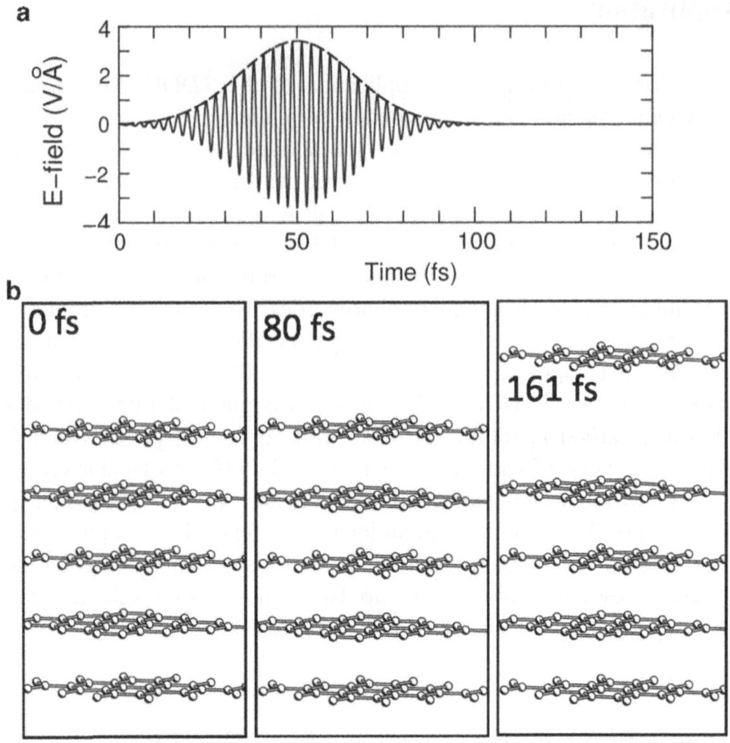

Fig. 2 (**a**) Pulse shape considered in current simulation having wavelength 800 nm and pulse width 45 fs (full width of half maximum). (**b**) Dynamics on the surface of graphite. *Small circles* are carbon atoms. See more details in [18]

During irradiation electronic excitation already occurs giving corresponding fluence (increased total energy given by Eq. (15)) as 87.9 mJ/cm^2. However, the ion dynamics is not significant at this moment due to large mass difference between electrons and ions. The growth of force field then finally let ions to move which makes ionic replacement apparent in later time. With shorter pulse (FWHM = 10 fs) having the same maximum intensity of the laser field, the dynamics is significantly slow down, still only the topmost layer is exfoliated.[1] The mechanism of exfoliation by femtosecond laser is supposed to be breakage of balance of charge redistribution among the layers as well as emitted electrons in vacuum due to laser irradiation. Yet further analysis would be needed.

[1]The result of shorter pulse with FWHM = 10 fs was revisited; the numerical data obtained in our former calculations [18] was carefully re-analyzed. We concluded that the top-layer exfoliation occurs as in the case with FWHM = 45 fs, but with slower speed. Meanwhile, the re-analysis of the data for FWHM = 45 fs [18] did not provide any update.

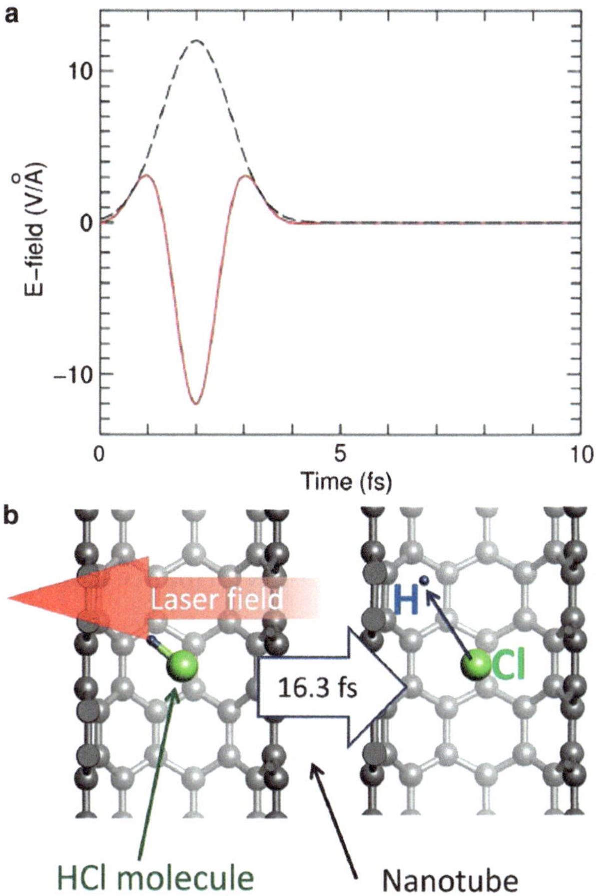

Fig. 3 (**a**) Pulse shape of the laser shot and (**b**) dynamics of a HCl molecule inside (8,0) nanotube. A *hatched arrow* in *left panel* denotes direction of light polarization. While a *solid arrow* in *right panel* denote trajectory of H atom detaching from Cl atom. More details are presented in [22]

3.2 *Pulse Induced Dynamics of Molecules Encapsulated Inside Carbon Nanotube*

As shown before, inter-layer interaction of bulk graphite can be broken by femtosecond laser shot. To break the stronger chemical bond, we need shorter pulse. Disintegration of finite systems (clusters and molecules) upon strong and short laser field was theoretically investigated [19, 20]. Here the system has been extended as molecules encapsulated in carbon nanotube (CNT) with periodic boundary conditions. As mentioned in beginning of this section, the optical penetration into CNT was of fundamental interest. According to our previous study [21], penetration of optical field into semiconducting single-walled CNT is feasible and even enhancement of the field takes place when the optical frequency is in resonance with excitation energy of the CNT. Motivated with this result, we have examined the feasibility of molecular disintegration of hydrochloride (HCl) molecule inside the (8,0) semiconducting CNT as a test case [22].

Figure 3 shows dynamics of HCl molecule encapsulated inside CNT. The pulse shape is asymmetric with respect to time so polarity of the optical electric field mostly direct only in one direction as time-average. The initial orientation of the HCl molecule is tilted to the CNT axis and spontaneous H emission is observed upon irradiation of the pulse. Meantime, the Cl atom remains immobile reflecting the mass-difference between these elements. The ejected H atom will gently collide to CNT inner wall and change its direction departing from the Cl atom. Such dynamics is sensitive to the molecular orientation. When the HCl molecular axis is perpendicular to CNT axis, the ejected H atom gives damage on the CNT inner wall creating single vacancy [22]. From the current result we can expect selective H-ejection from other kinds of molecules containing H atoms.

4 Some Requirements on High-Performance Computing

In this final section, some requirements on high-performance computer for performing the TDDFT-MD simulation are mentioned. Since TDDFT-MD simulation needs smaller interval of time by 10^{-3} of that of conventional *ab initio* MD simulation based on the Born-Oppenheimer approximation (BOA), the total number of time steps should be huge. The BOA-MD can reconfigure electron wave functions at every time-step since this formula solves *static* solutions. Meantime, the TDDFT-MD should update the wave function according to the numerical time-integration. By considering simulation time-scales targeted by BOA-MD and TDDFT-MD, the necessary computational task and accuracy is listed in Table 1.

Since required number of time-steps is higher and preservation of numerical accuracy is severer in TDDFT-MD simulations, one must be critical for the total energy conservation rule mentioned in Sect. 2.3 in performing the simulation.

Furthermore, it must be noted that the performing the Suzuki–Trotter split operator method [5] for integrating wave function along with the time-axis requires higher memory performance than static calculation. In static simulation, the most time-consuming operation is a product of the Kohn–Sham Hamiltonian and the Kohn–Sham orbitals like as

$$H_{KS}(\boldsymbol{r})\psi_n^{KS}(\boldsymbol{r}) \tag{17}$$

Since $H_{KS}(\boldsymbol{r})$ consists from many operators as shown in Eq. (5), we generalize to express $H_n^{KS}(\boldsymbol{r})$ as

$$H_{KS}(\boldsymbol{r}) \equiv H_{KS}^1(\boldsymbol{r}) + H_{KS}^2(\boldsymbol{r}) + H_{KS}^3(\boldsymbol{r}) + \cdots + H_{KS}^N(\boldsymbol{r}), \tag{18}$$

Table 1 Comparison in MD with static DFT and TDDFT

Method	Numerical error	Simulation time	Needed steps
BOA-MD[a]	Marginal	$\sim$10 ps	10^5
TDDFT MD[b]	Severe	500 fs $\sim$ 1 ps	$7 \times 10^5 \sim 10^6$

[a] MD with solving static DFT problem for electrons.
[b] MD with TDDFT time-evolution for electron wave functions.

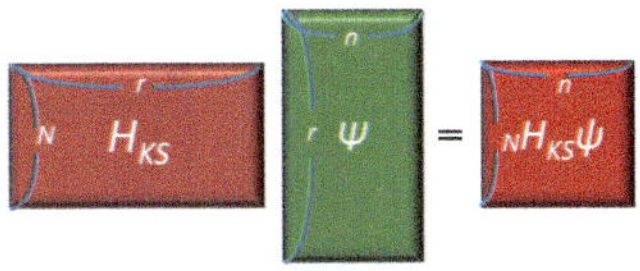

Fig. 4 Schematic showing products of $H_{KS}(\boldsymbol{r})$ and $\psi_m^{KS}(\boldsymbol{r})$ as matrix $\times$ matrix

where the number of operators N depends on the size of model coming from the non-local pseudopotentials shown in Eq. (4). So both $H_{KS}(\boldsymbol{r})$ and $\psi_n^{KS}(\boldsymbol{r})$ can be regard as matrixes whose suffices are discretized grid points of $\boldsymbol{r}$ and operator index N or band index n as illustrated by Fig. 4.

This matrix $\times$ matrix type operation can reduce memory access times by applying BLAS type algorisms and by blocking the size of matrices suitable to the size of on-tip memory in order to reduce data-transfer task between main memory and cache memory.

On the other hand, the time-evolution operator (second order with respect to $x = \delta t / i\hbar$) shown in Eq. (10) can be rewritten by inserting Eq. (18) as

$$e^{xH_{KS}(\boldsymbol{r})} = e^{\frac{x}{2}H^1_{KS}(\boldsymbol{r},t)} \left(\prod_{k=2}^{N-1} e^{\frac{x}{2}H^k_{KS}(\boldsymbol{r},t)} \right) e^{xH^N_{KS}(\boldsymbol{r},t)} \left(\prod_{k=N-1}^{2} e^{\frac{x}{2}H^k_{KS}(\boldsymbol{r},t)} \right) e^{\frac{x}{2}H^1_{KS}(\boldsymbol{r},t)}. \tag{19}$$

Contrary to Eq. (4) in which the sum of operators $H^k_{KS}(\boldsymbol{r})$ acts on matrix $\psi_n^{KS}(\boldsymbol{r})$, Eq. (19) requires sequential products of operators of $H^k_{KS}(\boldsymbol{r})$ on matrix $\psi_n^{KS}(\boldsymbol{r})$.

Indeed, operators $H^2_{KS}(\boldsymbol{r},t) \sim H^{N-1}_{KS}(\boldsymbol{r},t)$ are series of $V^{\tau,l}_{nl}(\boldsymbol{r}',\boldsymbol{r};t)$, that has a projection operator $\mid p_k(\boldsymbol{r},t) >< p_k(\boldsymbol{r}',t) \mid$ due to its separable form [23]. By using the exponential form of the projection operator [6],

$$e^{x|p_k(\boldsymbol{r},t)><p_k(\boldsymbol{r}',t)|} = 1 + \mid p_k(\boldsymbol{r},t) > C_k < p_k(\boldsymbol{r}',t) \mid, \tag{20}$$

with which C_k is a c-number as a function of x, we can illustrate multiple operation of $\mid p_k(\boldsymbol{r},t) >< p_k(\boldsymbol{r}',t) \mid$ to $\psi_n^{KS}(\boldsymbol{r},t)$ as Fig. 5. So vector $\times$ matrix operations are repeated.

It is therefore impossible to simply apply the BLAS type tool like the case of matrix $\times$ matrix operation and thus the TDDFT calculation requires higher memory performance than static DFT calculation. Developments in new algorisms to save

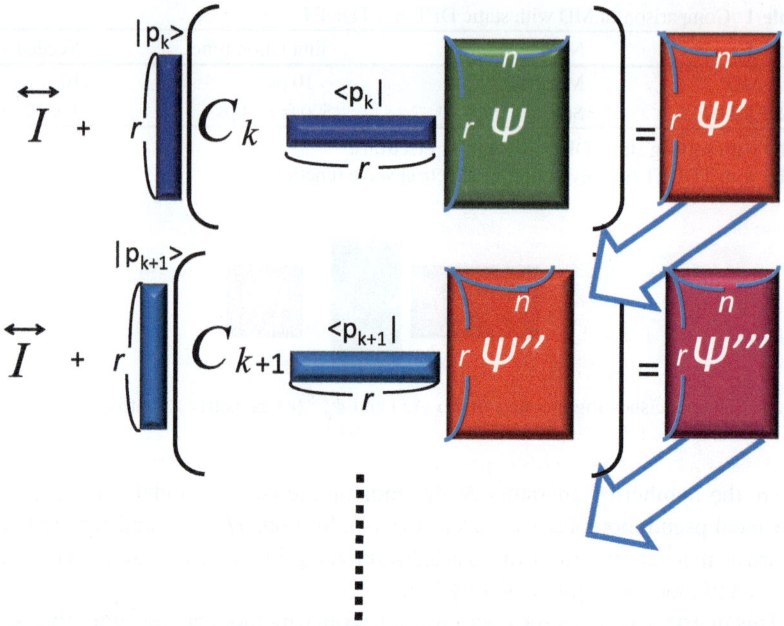

Fig. 5 Schematic showing series of products of $H_{KS}^{k}(\boldsymbol{r},\boldsymbol{r}',t)$, which are non-local parts, to $\psi_n^{KS}(\boldsymbol{r})$ as vector × matrix operations

memory performance or new computer architecture to achieve efficient memory performance would be required.

5 Summary and Conclusion

Computational method to treat electron–ion dynamics in condensed matters under irradiation of intense and short laser shot has been presented. Time-dependent density functional theory requires numerical solution of the time-dependent Schrödinger equation (time-dependent Kohn–Sham equation) that can practically possible by applying the Suzuki–Trotter formula. Application of this computational scheme to laser induced graphene exfoliation from graphite and molecular disintegration inside CNT are demonstrated. This simulation will stimulate experimental exploration using the femtosecond laser but also requires development in algorisms and computer architecture.

Acknowledgements All calculations shown in this work were done by using the Earth Simulator. These works were done under collaboration with D. Tománek, H. Zhang and A. Rubio.

References

1. Shimotsuma Y, Hirao K, Kazansky P. G, and Qiu J (2005): Three-Dimensional Micro- and Nano-Fabrication in Transparent Materials by Femtosecond Laser. Jpn. J. Appl. Phys. **44**, 4735–4748
2. Kohn W, and Sham LJ (1965): Self-Consistent Equations Includinf Exchange and Correlation. Phys. Rev. **140**: A1133–A1138
3. Hohenberg P and Kohn W (1964): Inhomogeneous Electron Gas. Phys. Rev. **136**: B864–B871
4. Runge E and Gross EKU (1984): Density-Functional Theory for Time-Dependent System. Phys. Rev. Lett. **52**:997–1000.
5. Suzuki M (1992):General Nonsymmetric Higher-Order Decomposition of Exponential Operators and Symplectic Integrators. J. Phys. Soc. Jpn, **61**:3015–3019
6. Sugino O and Miyamoto Y (1999): Density-functional approach to electron dynsmics: Stable simulation under a self-consisten field. Phys. Rev.B**59**:2579–2586; (2002) Phys. Rev. B**66**:089901 (E).
7. Ehrenfest P (1927).: Bemerkung über die angenäherte Gültigkeit der klassischen Mechanik innerhalb der Quantenmechanik. Z. Phys. **45**:455–457
8. Miyamoto Y and Zhang H (2008): Testing the numerical stability of time-dependent density functional simulations using the Suzuki–Trotter formula. Phys. Rev. B**77**:165123-1–165123-5
9. Novoselov KS, Geim AK, Morozov SV, Jiang D, Zhang Y, Dubonos SV, Grigorieva IV, and Firsov AA: (2004) Electric Field Effect in Atomically Thin Carbon Films, Science **306**:666–669
10. Reina A, Jia X, Ho J, Nezich D, Son H, Bulovic V, Dresselhaus MS, and Kong J (2009): Large Area, Few-Layer Graphene Films on Arbitrary Substrates by Chemical Vapor Deposition, Nano Lett. **9**:30–35
11. Li X, Gai W, An J, Kim S, Nah J, Yang D, Piner R, Velamakanni A, Jun I, Tutuc E, Banerjee SK, Colombo L, and Ruoff RS (2009): Large-Area Synthesis of High-Quality and Uniform Graphene Films on Copper Foils, Science **324**:1312
12. Tromp RM and Hannon JB (2009): Thermodynamics and Kinetic of Graphene Growth on SiC(0001), Phys. Rev. Lett. **102**:106104-1–106104-4
13. Carbone F, Baum P, Rudolf P, and Zewail AH (2008): Structural Preablation Dynamics of Graphene Observed by Ultrafast Electron Crystallography, Phys. Rev. Lett. **100**:035501-1–035501-4
14. Raman RK, Murooka Y, Ruan CY, Yang T, Berber S and Tománek D (2008): Direct Observation of Optically Induced Transient Structures in Graphite Using Ultrafast Electron Crystallography, Phys. Rev. Lett. **101**:077401-1–077401-4.
15. Jeschke HO, Garcia ME and Bennemann KH (2001): Theory for the Ultrafast Ablation of Graphite Films, Phys. Rev. Lett. **87**:015003-1–015003-4
16. Ajyayam PM and Iijima S (1993): Capillarity-induced filling of carbon nanotubes. Nature, **361**:333–334
17. Yanagi K, Iakoubovskii K, Kazaoui S, Minami N, Maniwa Y, Miyata Y, and Kataura H (2006): Light-harvesting function of β-carotene inside carbon nanotubes, Phys. Rev. B**74**:155420-1–155420-6
18. Miyamoto Y, Zhang H, and Tománek D (2010): Photoexfoliation of Graphene from Graphite: An *Ab Initio* Study, Phys. Rev. Lett. **104**:208302-1–208302-4
19. Castro A, Marques MAL, Alonso JA, Bertsch GF, and Rubio A (2004): Excited states dynamics in time-dependent density functional theory, Eur. Phys. D**28**:211–218
20. Taguchi K, Haruyama J, and Watanabe K (2009): Laser-Driven Molecular Dissociation: Time-Dependent Density Functional Theory and Molecular Dynamics Simulations, J. Phys. Soc. Jpn, **78**:0947071-1–094707-6
21. Zhang H and Miyamoto Y (2009): Modulation of alternating electric field inside photoexcited carbon nanotubes, Appl. Phys. Lett., **95**:053109-1–053109-3

22. Miyamoto Y, Zhang H, and Rubio A (2010): First-Principles Simulation of Chemical Reactions in an HCl Molecule Embedded inside a C or BN Nanotube Induced by Ultrafast Laser Pulses, Phys. Rev. Lett., **105**:248301-1–248301-4
23. Kleinman L and Bylander DM (1982): Efficacious Form for Model Pseudopotentials, Phys. Rev. Lett., **48**:1424–1428

Numerical Investigation of Nano-Material Processing by Thermal Plasma Flows

Masaya Shigeta

Abstract The high performance of supercomputing systems has made it feasible to clarify multi-scale physics of nano-material processes in thermal plasma environments with thermofluidic and electromagnetic interactions.

In this chapter, two challenges related to the thermal plasma processing of nanoparticle fabrication are presented as new applications of supercomputing. The growth process of titanium boride nanoparticles from the precursory binary vapors of titanium and boron is computed using a unique mathematical model. In consequence, the collective and simultaneous growth behavior through nucleation, co-condensation and coagulation is clarified. The 3-D complex structure of the thermofluid field in/around an argon inductively coupled thermal plasma (ICTP) has also been revealed. The higher-temperature region has larger vortices, whereas the lower-temperature flow forms smaller eddies. Because a high-temperature plasma has a high electrical conductivity, the Lorentz forces are generated there; and consequently recirculating zones are produced. In addition, it is also clarified that turbulent and laminar regions co-exist and form a complicated flow field in the ICTP torch.

1 Introduction

Efficient production of nanoparticles is intensively expected for numerous applications in industrial, biomedical, and environmental purification processes because nanoparticles exhibit unique electronic, optical, and catalytic properties which are different from those of larger particles of micrometer size or bulk materials. However, the synthesis of functional nanoparticles of ceramics and alloys seems

M. Shigeta (✉)
Tohoku University, 6-6-01 Aramaki-Aoba, Aoba-ku, Sendai 980-8579, Japan
e-mail: shigeta@fluid.mech.tohoku.ac.jp

M. Resch et al. (eds.), *Sustained Simulation Performance 2012*,
DOI 10.1007/978-3-642-32454-3_14,

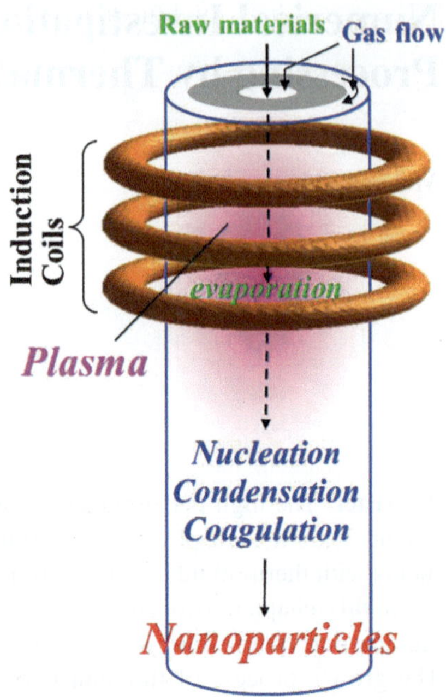

Fig. 1 Nanoparticle synthesis system utilizing an ICTP flow

to be practically arduous by conventional methods because their raw materials have too high melting points to be decomposed. Even combustion processes cannot generate a sufficiently high temperature to vaporize the raw materials; furthermore, combustion requires an oxidation atmosphere, which is undesirable for synthesis of nanoparticles and which produces contamination from combustion products (CO_2, H_2O, etc.).

To overcome such a problem, thermal plasmas have been anticipated as a promising tool for efficient fabrication of nanoparticles [1] because thermal plasmas offer a distinctive thermofluid field involving a higher temperature, high chemical reactivity and variable properties. Moreover, thermal plasmas are controllable by external electromagnetic fields [2–4]. In particular, an inductively coupled thermal plasma (ICTP) illustrated in Fig. 1, which is produced without internal electrodes, inherently has a contamination-free large plasma field. By virtue of such a field, a large amount of raw materials, even that with high melting/boiling points, are vaporized completely. The vapors of the raw materials are transported downstream with the plasma flow to the plasma's tail region where the temperature decreases rapidly; and consequently the vapors become highly supersaturated. Because the supersaturated state is unstable, the vapors change their phases quickly to very small particles like fog or smoke through the collective and simultaneous formation through nucleation, condensation and coagulation among the particles themselves. In consequence, nanoparticles are mass-produced at a high rate.

However, it is still difficult to investigate the formation mechanism of functional nanoparticles of ceramics and alloys generated in/around an ICTP because the process involves remarkably severe and intricate heat/mass transfer associated with phase conversions in a few tens of milliseconds in a complex thermofluid field interacting with an induced electromagnetic field. Experimental approaches using a direct measurement or observation are currently impossible owing to technological limitations, whereas numerical studies have struggled with the mathematical formulation to express the intricate multi-scale physics and the shortage of computational resources to obtain practically meaningful solutions.

Meanwhile, the recent progress of supercomputing systems is noticeable, which has made it possible to simulate multi-scale and complicated phenomena with sufficient accuracy: for example, for global meteorological simulations [5] and aerodynamic optimization in airplane designing [6]. Moreover, the first principle simulations using large-scale supercomputing resources have also obtained many significant results [7]. Using the present supercomputing systems, numerical investigations of the nano-material processing by thermal plasma flows will be feasible as well.

This chapter introduces two challenges related to the thermal plasma processing of nanoparticle fabrication as new applications of supercomputing. Section 2 presents a computational clarification of a binary growth process of functional ceramic nanoparticles; and Sect. 3 demonstrates a time-dependent 3-D large eddy simulation (LES) of the complex thermofluid field in/around an ICTP flow.

2 Binary Growth of Functional Nanoparticles

2.1 Model Description

The vapor phase syntheses of binary functional nanoparticles can be computed using a unique mathematical model that was recently developed by the author [8, 9]. The model describes a collective and simultaneous growth process of two-component nanoparticles in a binary vapor system with the following assumptions: (i) nanoparticles are spherical; (ii) inertia of nanoparticle is negligible; (iii) the temperature of nanoparticles is identical to that of the bulk gas surrounding them; (iv) heat generated by condensation and the electric charge of nanoparticles are neglected; and (v) the material vapors are regarded as an ideal gas. To treat the particle size and composition during the growth, the model introduces the particle size-composition distribution (PSCD) on the basis of two-directional nodal discretization [10]. Using the PSCD, the net production rate of nanoparticles having the volume v_k and the content of the second material x_n is written by the increment of the number density N during the infinitesimal time Δt as:

$$
\begin{aligned}
\frac{\Delta N_{k,n}}{\Delta t} &= J_{binary}\xi_k^{(nucl)}\psi_n^{(nucl)} \\
&+ \sum_i \sum_l \frac{\left(\xi_{i,l,k}^{(cond)}\psi_{i,l,n}^{(cond)} - \delta_{i,k}\delta_{l,n}\right) N_{i,l}}{\Delta t} \\
&+ \frac{1}{2}\sum_i \sum_j \sum_l \sum_m \xi_{i,j,k}^{(coag)}\psi_{i,j,l,m,n}^{(coag)}\beta_{i,j,l,m}N_{i,l}N_{j,m} \\
&- N_{k,n}\sum_i \sum_l \beta_{i,k,l,n}N_{i,l}. \qquad (1)
\end{aligned}
$$

Here, ξ and ψ denote the splitting operators for the size and composition, respectively. δ is the Kronecker's delta. The first term on the right-hand side represents the contributions of binary homogeneous nucleation. The second term means the production rate caused by vapor condensation on the particles having v_i and x_l. The third and fourth terms express the gain and loss by coagulation among nanoparticles. β is the collision frequency function for nanoparticles resulting from Brownian motion [11]. Subscripts i and j denote the node numbers for size, whereas subscripts l and m signify those for composition. To obtain the homogeneous nucleation rate J_{binary} for a binary system, the theoretical formula derived by Wyslouzil and Wilemski [12] is used.

When the growth rate of nanoparticles by heterogeneous condensation of the vapor of material M, the following formula considering the rarefied gas effect correction is used to calculate the volume increment $v_{(M)i,l}$ during the infinitesimal time increment Δt:

$$
\begin{aligned}
\frac{\Delta v_{(M)i,l}}{\Delta t} &= 2\pi d_i D_{vap(M)} v_{vap(M)}\left(N_{vap(M)} - \bar{N}_{S(M)i,l}\right) \\
&\times \left[\frac{0.75\alpha_{(M)}(1+\mathrm{Kn}_i)}{0.75\alpha_{(M)} + 0.283\alpha_{(M)}\mathrm{Kn}_i + \mathrm{Kn}_i + {\mathrm{Kn}_i}^2}\right], \qquad (2)
\end{aligned}
$$

where d is the diameter, D is the diffusion coefficient, and Kn is the Knudsen number. The subscript vap denotes vapor. α represents the accommodation coefficient which has a value of 0.1 here. $\bar{N}_S$ means the saturated vapor concentration considering the effects of material mixture and surface curvature [13]. Note that the nanoparticles are allowed to grow by condensation only when the free energy gradients for particle formation, W, in a binary system is negative or zero:

$$\frac{\partial W}{\partial n_{(M)}} \leq 0 \qquad (3)$$

where, $n_{(M)}$ represents the number of monomers of material M in a nanoparticle. W is composed of the chemical potentials and the surface energy.

The population balance equations of the material vapors are also computed simultaneously because the number densities of the material vapors crucially affect the growth process:

$$\frac{\Delta N_{vap(M)}}{\Delta t} = -\sum_{k} J_{binary}\xi_k^{(nucl)} n_{(M)}^* - \sum_{i}\sum_{l} \frac{N_{i,l}\Delta v_{(M)i,l}}{v_{vap(M)}\Delta t}, \quad (4)$$

where $n_{(M)}^*$ represents the number of monomers of material M composing a stable nucleus.

In the computation, the melting point depressions due to nano-scale size and the mixture effect are considered [14]. It is assumed that the nanoparticles with the temperature lower than their melting point cannot grow by coagulation.

2.2 Computational Conditions

The synthesis of titanium boride nanoparticles is selected as a target process. Nanoparticles of titanium borides exhibit several advantages: high melting points, strengths, hardnesses, durabilities, wear resistances, and electrical conductivities, and besides low work functions. Hence, those nanoparticles are expected to be applied to electromagnetic shielding, wear-resistant coatings, and solar control windows [15, 16]. In addition to the strong demands in industry, those nanoparticles have attracted a plenty of interests of physicists and chemists as well as engineers because their collective and simultaneous growth from the vapor phases is a complicatedly interacting process and the fabrication of them with well-controlled sizes and compositions is still very difficult. Those problems are fundamentally attributed to the large difference of the saturation vapor pressures of titanium and boron (e.g. Ti/B $= 10^2$).

Currently, only the model described in Sect. 2.1 seems to be able to analyze the detailed mechanism of the collective and simultaneous growth of titanium boride nanoparticles.

The computation is demonstrated under a typical condition of the synthesis using a thermal plasma as follows. Coarse precursory powders of titanium and boron are injected with the feed rate of 0.5 g/min in the ratio of Ti:B $=$ 3:1 into the thermal plasma. They are vaporized immediately in the high-enthalpy plasma and the vapors are transported downstream with the decrease in their temperature. The computation begins from this situation in which the vapors are about to be supersaturated. A one-dimensional profile of a typical bulk gas condition at the plasma tail is used, which was obtained by a preliminary calculation based on electromagnetic thermofluid dynamics [8]. There, the temperature decreases monotonically. Moreover, the decreasing rate changes from 1.5×10^5 to 1.0×10^4 K/s. In response to that decreasing rate, the time increment Δt for the computation is increased from 5.0×10^{-6} to 2.0×10^{-5} s. With this time increment, 19,422 time steps

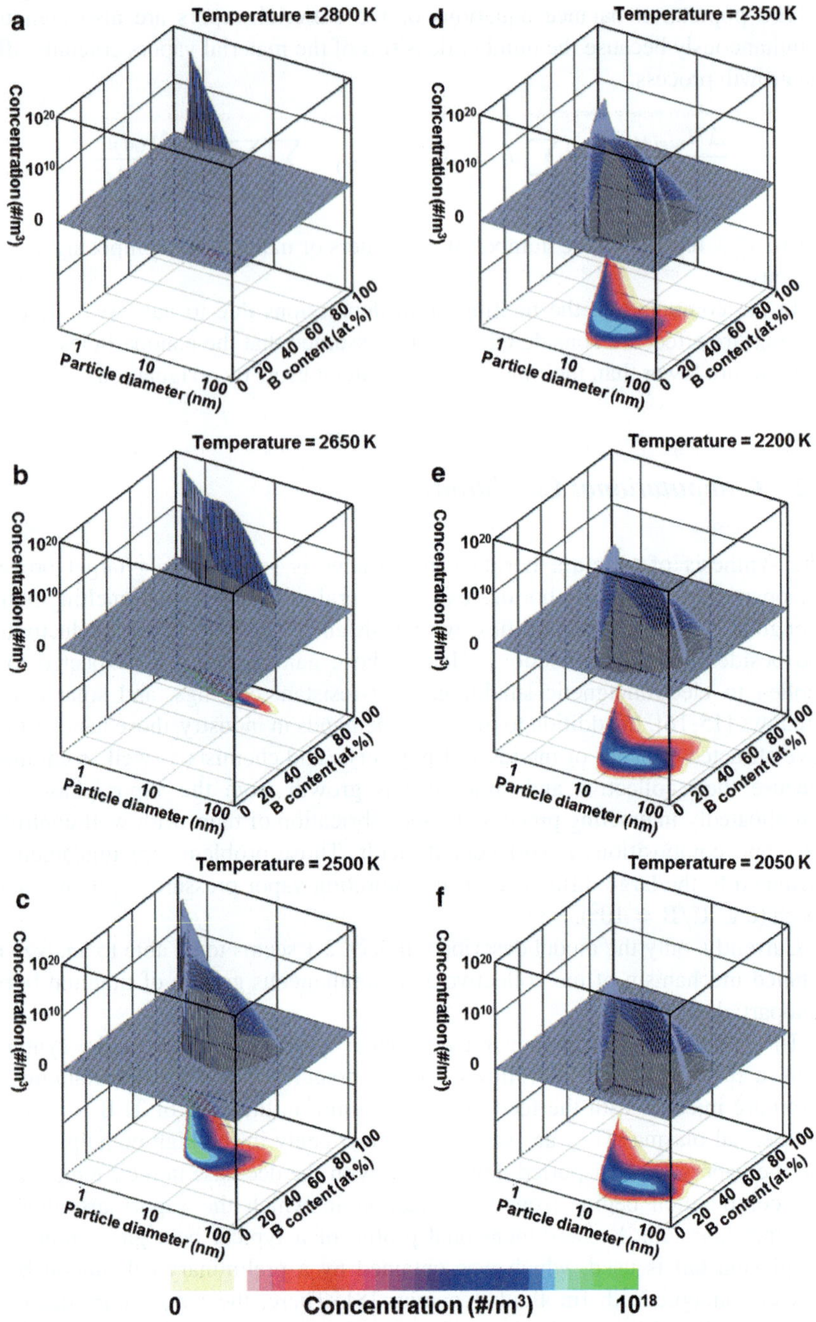

Fig. 2 Collective and simultaneous growth behavior of titanium boride nanoparticles under a temperature-decreasing condition

were required until the growth process was completed. As described at Sect. 2.1, the mathematical model embeds the discrete nodes in the size-composition space. Here, 101×101 nodes were placed in the size and composition directions, respectively; in consequence, a wide range of the particle diameters from sub-nanometers to 159 nm was covered. The material properties of titanium and boron were obtained from [17].

The computation was performed with the supercomputer at Cyberscience center of Tohoku University (Express5800, NEC, Japan) using 32 CPU (Threads). The solver was coded in FORTRAN and parallelized by OpenMP.

2.3 Numerical Results

Figure 2 shows the time evolution of the PSCD in the Ti-B binary system, which expresses the collective and simultaneous growth of titanium boride nanoparticles. While the temperature decreases around 2,800 K, boron vapor reaches a supersaturated state at a higher temperature than titanium vapor. The supersaturated boron vapor starts to generate stable nuclei which are the embryos of nanoparticles; and the nuclei grow up to boron-rich nanoparticles. Immediately, the vapors of boron and titanium co-condense on the existing nanoparticles, where the boron vapor has a higher rate of condensation than the titanium vapor because the saturation pressure of boron is much lower than that of titanium [18, 19]. Following the consumption of boron vapor, titanium vapor is consumed with a high rate of condensation. During this process, coagulation among the nanoparticles also takes place simultaneously. As a result of this growth behavior, the Ti-B system produces nanoparticles having wide ranges of the size and the boron content because of the time lag of co-condensation between boron and titanium. Figure 2 also tells that the majority of the nanoparticles have the diameters around 20 nm and the boron content of 25 at.%. It is noteworthy that the other smaller or larger nanoparticles have larger contents of boron. These features are determined by the balance of the particle size and the condensation rate which is a function of the particle size as described in Eq. (2).

3 Time-Dependent 3-D Simulation of an ICTP Flow

3.1 Model Description

An inductively coupled thermal plasma (ICTP) flow is described by thermofluid dynamics coupled with electromagnetics. There, the entire flow field in which the plasma at a high temperature and a cold gas at a room temperature co-exist must be treated at the same time, which makes thermal plasma simulations very arduous. The widely ranging temperature from 300 to 12,000 K leads to large spatial variations of the transport properties with several orders of magnitude. Moreover, the density at a room-temperature region is approximately 50 times of that at a high-temperature plasma. Meanwhile, the Mach numbers in/around the plasma are estimated to range from 0.003 to 0.015. Hence, a thermal plasma is treated as an incompressible flow with the density as a temperature-dependent variable to obtain the solution in a practical time-scale.

To formulate the governing equations of an argon inductively coupled thermal plasma (ICTP) generated at atmospheric pressure, the following assumptions are validly adopted: (i) local thermodynamic equilibrium, (ii) electrically neutral, (iii) optically thin, (iv) negligible displacement current, and (v) negligible Hall effect.

The thermofluidic motion of an ICTP flow is described by the simultaneous conservation equations of mass, momentum and energy, associated with the Maxwell's equation to determine the electromagnetic field induced in and around the plasma:

$$\frac{\partial \rho}{\partial t} + \frac{\partial}{\partial x_j}\left(\rho u_j\right) = 0, \tag{5}$$

$$\frac{\partial}{\partial t}\left(\rho u_i\right) + \frac{\partial}{\partial x_j}\left(\rho u_j u_i\right) = -\frac{\partial P}{\partial x_i} + \frac{\partial \tau_{ji}^{GS}}{\partial x_j} + \frac{\partial \tau_{ji}^{SGS}}{\partial x_j} + \frac{1}{2}\sigma\mu_0 \mathrm{Real}\left(\epsilon_{ijk} E_j \tilde{H}_k\right) + \rho g_i, \tag{6}$$

$$\frac{\partial}{\partial t}\left(\rho h\right) + \frac{\partial}{\partial x_j}\left(\rho u_j h\right) = -\frac{\partial q_j^{GS}}{\partial x_j} + \tau_{ji}^{GS} S_{ij} + \epsilon^{SGS} - \frac{\partial q_j^{SGS}}{\partial x_j} + \frac{1}{2}\sigma E_i \tilde{E}_i - R, \tag{7}$$

and

$$\mu_0 \sigma \frac{\partial A_l}{\partial t} = \frac{\partial}{\partial x_k}\left(\frac{\partial A_l}{\partial x_k}\right) - i\mu_0\sigma\omega A_l. \tag{8}$$

Here, ρ is the density, t is the time, u_i is the velocity vector, x_i is the position vector, P is the pressure, τ_{ij} is the stress tensor, σ is the electrical conductivity, μ_0 is the permeability in vacuum, ϵ_{ijk} is the Eddington's epsilon, E_i is the electric field vector, H_i is the magnetic field vector, g_i is the gravitational acceleration vector, h is the enthalpy, q_i is the heat flux vector, ϵ is the viscous dissipation rate, R is the radiation loss, A_i is the vector potential, and ω is the angular frequency. i is the imaginary unit, $i = \sqrt{-1}$. The subscripts i, j, k and l denote the directions. The superscripts GS and SGS signify the grid scale and the sub-grid scale, respectively. S_{ij} is the velocity strain tensor defined as

$$S_{ij} = \frac{1}{2}\left(\frac{\partial u_j}{\partial x_i} + \frac{\partial u_i}{\partial x_j}\right). \tag{9}$$

The fourth term on the right-hand side in Eq. (6) expresses the effective value of the Lorentz force caused by the induced electromagnetic field. The tilde (˜) means the complex conjugate. In addition, the gravitational force is also considered because the effect of gravity can be significant in the plasma torch where low-density fluid with high-temperature and high-density fluid with low-temperature co-exist. The fifth term on the right-hand side in Eq. (7) describes the effective value of the Joule heating, whereas the radiation loss at the sixth term is taken into account because it offers a considerable cooling effect on plasma. The terms at the sub-grid scale are locally determined by the coherent structure Smagorinsky model [20].

The electric field and the magnetic field are obtained from the vector potential using the following relations:

$$E_l = -i\omega A_l - \frac{\partial A_l}{\partial t} \tag{10}$$

and

$$\mu_0 H_j = \epsilon_{jkl} \frac{\partial A_l}{\partial x_k}. \tag{11}$$

3.2 Computational Conditions

In this section, a time-dependent 3-D simulation of an argon ICTP is demonstrated to clarify the complex flow structure during plasma discharge. In Fig. 1, the central carrier gas is not injected; however, only the swirl sheath gas at 300 K is injected at 23 Sl/min from the slit at the torch top. The torch wall is made of quartz with a thickness of 2 mm and the outside of the wall is water-cooled at 300 K constant. An induction coil which has a spiral shape is set around the torch. With the coil, the electromagnetic field is applied with the power of 6 kW and the frequency of 4 MHz.

A quasi-parabolic velocity profile of a fully-developed laminar flow in an annulus [21] is given to the injected flow. At the torch wall, the heat transfer from the fluid to the outside of the wall is taken into account. For the induced electromagnetic field, the contributions from both the coil current and the current induced in the plasma are considered on the boundary condition of the vector potential at the wall, which can be derived from the Biot–Savart law [22].

The computation requires the temperature-dependent data of the properties of an argon thermal plasma up to 12,000 K. The data of the density, the viscosity, the thermal conductivity, the specific heat, and the electrical conductivity were obtained from [23]. In addition, the data curve of the radiation loss was obtained from [24].

The computational domain was discretized into 256 × 101 × 101 control volumes by the Finite Volume Method. The second-order central differencing scheme combined with the first-order upwind differencing scheme in the ratio of 9:1 is applied to the convection terms to capture multi-scale vortices. For the time marching, the second-order Adams–Bashforth scheme is applied to the transient terms. The diffusion terms and the source terms are discretized by the second-order central-differencing scheme. Implementing these schemes, the governing equations are solved by the PISO (Pressure-Implicit with Splitting Operators) algorithm [25, 26]. The axisymmetric profile is given as the initial condition (time $t = 0$).

To obtain the numerical results for a real time period of 1.0 s, the computation took approximately 20 days for 2,000 time steps with a time interval Δt of 0.5 ms, using the supercomputer at Cyberscience center of Tohoku University (SX-9, NEC, Japan). The solver was also coded in FORTRAN and parallelized by OpenMP to use 16 CPU (Threads). The computation was performed with the average vector length of 244.83 and the vector operation ratio of 99.69%, using the memory of 5.312 GB.

3.3 Numerical Results

Figure 3 shows the snapshots of the instantaneous thermofluid field in the ICTP torch; Fig. 3a–c presents the coherent vortex structure visualized by the isosurfaces for $Q = 0.5$; and Fig. 3d–f portray the isosurfaces for 7,000 K and the streamlines. Note that Q is the second invariant of the velocity gradient tensor normalized by an arbitrary representative speed of 10 m/s and the torch diameter of 50 mm.

It is revealed that the higher-temperature region in/around the plasma has larger vortices, whereas the lower-temperature flow forms smaller eddies. The largest vortex structure around 10,000 K appears to stay in the plasma region because of the Lorentz force. On the other hand, the small cold eddies are observed near the top and side walls of the torch. This is reasonable because a low-temperature fluid has a much lower viscosity than the high-temperature plasma. That is, the cold gas flow apparently has a higher local Reynolds number, where smaller eddies tend to be generated and survive. The streamlines show a 3-D complex structure with a noticeable recirculating zone in the plasma. Because a high-temperature plasma has a high electrical conductivity, the Lorentz forces are generated there and they drive fluid motions. In consequence, recirculating flows are produced in/around the plasma.

Figure 4 shows the characteristics of the flow field at a certain moment. In Fig. 4a, the isosurfaces for $Q = 1.0$ are portrayed with colors indicating the absolute value of the normalized helicity density. The color profiles on the coherent vortices tell that the vectors of the velocity and the vorticity are oriented in the same directions only on a few vortex structures. Figure 4b depicts the distribution of the viscosity ratio which is defined as the ratio of the turbulent eddy viscosity to the molecular viscosity. That is, the regions having a higher viscosity ratio are more turbulent, whereas the regions having a lower viscosity ratio are more laminar. Figure 4b indicates that the flow in an ICTP torch is a very complicated field where turbulent regions and laminar regions co-exist.

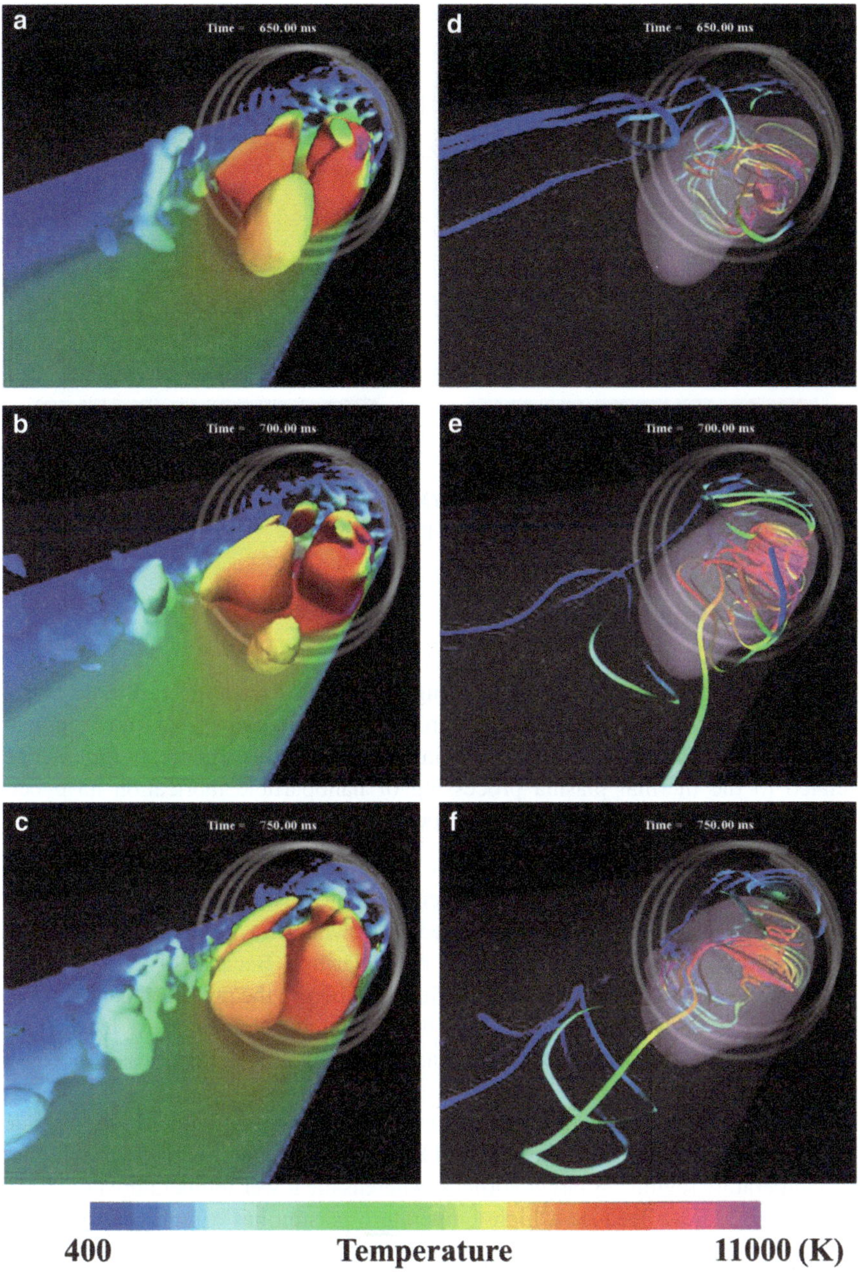

Fig. 3 Instantaneous thermofluid field in ICTP torch: (**a**)–(**c**) coherent vortex structure visualized by isosurfaces for $Q = 0.5$, and (**d**)–(**f**) isosurfaces for 7,000 K and streamlines

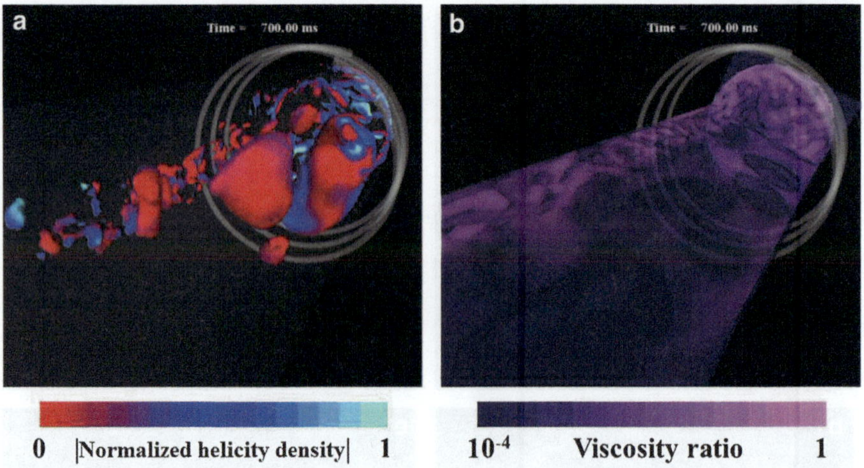

Fig. 4 Flow characteristics: (**a**) isosurfaces for $Q = 1.0$ with colors indicating normalized helicity density, and (**b**) viscosity ratio distribution on representative cross sections

4 Concluding Remarks

The high performance of supercomputing systems has made it feasible to clarify multi-scale physics of nano-material processes in thermal plasma environments with thermofluidic and electromagnetic interactions. In this chapter, two challenges related to the thermal plasma processing of nanoparticle fabrication have been presented as the latest applications of supercomputing.

1. The growth process of titanium boride nanoparticles from the precursory binary vapors of titanium and boron has been computed using a unique mathematical model. The collective and simultaneous growth behavior through nucleation, co-condensation and coagulation has been clarified. Because of the time lag of binary co-condensation between titanium and boron, the produced nanoparticles have wide ranges of the size and the boron content. The nanoparticles deviated from the mean-size nanoparticles tend to exhibit larger fractions of boron in them.
2. The 3-D complex structure of the thermofluid field in/around an argon inductively coupled thermal plasma (ICTP) has also been revealed. The higher-temperature region has larger vortices, whereas the lower-temperature flow forms smaller eddies. Because a high-temperature plasma has a high electrical conductivity, the Lorentz forces are generated there; and consequently recirculating zones are produced. In addition, it has also been clarified that turbulent and laminar regions co-exist and form a complicated flow field in the ICTP torch.

Acknowledgements This work was partly supported by the Japan Society for the Promotion of Science Grant-in-Aid for Scientific Research (C) (Grant No. 23560182); the results of this study were obtained using the supercomputing resources of Cyberscience Center, Tohoku University. The author expresses his gratitude to Dr. Yu Fukunishi and Dr. Seiichiro Izawa of Tohoku University for valuable advices about the vortex identification and to Dr. Takayuki Watanabe of Tokyo Institute of Technology for the experimental insight of titanium boride nanoparticle synthesis. In addition, the author would like to thank Mr. Takashi Soga of NEC System Technology, Ltd. and Mr. Takeshi Yamashita of Tohoku University for improving the solver code.

References

1. Shigeta, M., Murphy, A. B.: Thermal plasmas for nanofabrication. J. Phys. D: Appl. Phys. **44**, 174025 (16 pp) (2011)
2. Sato, T., Shigeta, M., Kato, D., Nishiyama, H.: Mixing and magnetic effects on a nonequilibrium argon plasma jet. Int. J. Thermal Sci. **40**, 273–278 (2001)
3. Shigeta, M., Sato, T., Nishiyama, H.: Computational experiment of a particle-laden RF inductively coupled plasma with seeded potassium vapor. Int. J. Heat Mass Trans. **47**, 707–716 (2004)
4. Shigeta, M., Nishiyama, H.: Numerical Analysis of Metallic Nanoparticle Synthesis using RF Inductively Coupled Plasma Flows. Trans. ASME, J. Heat Trans. **127**, 1222–1230 (2005)
5. Takahashi, K. et al.: World-highest resolution global atmospheric model and its performance on the Earth Simulator. Proc. SC '11 State of the Practice Reports Article No. 21 (12 pp) (2011)
6. Sasaki, D., Nakahashi, K.: Aerodynamic optimization of an over-the-wing-nacelle-mount configuration. Modelling Sim. Eng. **2011**, Article ID 293078 (13 pp) (2011)
7. Zhang, H., Miyamoto, Y.: Graphene production by laser shot on graphene oxide: An ab initio prediction. Phys. Rev. B **85**, 033402 (4 pp) (2012)
8. Shigeta, M., Watanabe, T.: Growth model of binary alloy nanopowders for thermal plasma synthesis. J. Appl. Phys. **108**, 043306 (15 pp) (2010)
9. Cheng, Y., Shigeta, M., Choi, S., Watanabe, T.: Formation mechanism of titanium boride nanoparticles by RF induction thermal plasma. Chem. Eng. J. **183**, 483–491 (2012)
10. Shigeta, M., Watanabe, T.: Two-directional nodal model for co-condensation growth of multicomponent nanoparticles in thermal plasma processing. J. Therm. Spray Technol. **18**, 1022–1037 (2009)
11. Seinfeld, J. H., Pandis, S.N.: Atmospheric Chemistry and Physics, From Air Pollution to Climate Change, pp. 660–661. Wiley, New York (1998)
12. Wyslouzil, B. E., Wilemski, G.: Binary nucleation kinetics. II. Numerical solution of the birth-death equations. J. Chem. Phys. **103**, 1137–1151 (1995)
13. Vesala, T., Kulmala, M., Rudolf, R., Vrtala, A., Wagner, P. E.: Models for condensational growth and evaporation of binary aerosol particles. J. Aerosol Sci. **28**, 565–598 (1997)
14. Wautelet, M., Dauchot, J. P., Hecq, M.: Phase diagrams of small particles of binary systems: a theoretical approach. Nanotechnology **11**, 6–9 (2000)
15. Lundstrom, T.: Structure, defects and properties of some refractory boride. Pure Appl. Chem. **57**, 1383–1390 (1985)
16. Munro, R. G.: Material properties of titanium diboride. J. Res. Natl. Stand. Technol. **105**, 709–720 (2002)
17. Japan Institute of Metals: Metal Data Book, Maruzen, Tokyo (1993)
18. Turov, Y. V., Khusid, B. M., Voroshnin, L. G., Khina, B. B., Kozlovskii, I. L.: Gas transport processes in sintering of an iron-boron carbide powder composite. Powder Metallurgy and Metal Ceramics **28**, 618–622 (1990)
19. Angelino, G., Angelino, L., Sirignano, W. A.; Modern Research Topics in Aerospace Propulsion: In Honor of Corrado Casci, pp. 59. New York, USA (1991)
20. Kobayashi, H.: The subgrid-scale models based on coherent structures for rotating homogeneous turbulence and turbulent channel flow. Phys. Fluid **17**, 045104 (12 pp) (2005)

21. Berman, A. S.: Laminar Flow in an Annulus with Porous Walls. J. Appl. Phys. **29**, 71–75 (1958)
22. Mostaghimi, J., Boulos, M. I.: Two-dimensional electromagnetic field effects in induction plasma modelling. Plasma Chem. Plasma Proc. **9**, 25–44 (1989)
23. Atsuchi, N., Shigeta, M., Watanabe, T.: Modeling of non-equilibrium argon-oxygen induction plasmas under atmospheric pressure. Int. J. Heat Mass Trans. **49**, 1073–1082 (2006)
24. Menart, J., Lin, L.: Numerical study of high-intensity free-burning arc. J. Thermophys. Heat Trans. **12**, 500–506 (1998)
25. Issa, R. I.: Silution of Implicitly Discretised Fluid Flow Equations by Operator-Splitting. J. Computational Phys. **62**, 40–65 (1985)
26. Oliveira, P. J., Issa, R. I.: An Improved PISO Algorithm for the Computation of Buoyancy-Driven Flows. Numerical Heat Trans. B **40**, 473–493 (2001)

Batch number: 10585138

Printed by Printforce, the Netherlands